...erstellar dust grain: diameter 4×10^{-5} inch

Blue light wavelength: 1.9×10^{-5} inch

Bacterium: diameter 4×10^{-5} inch

...ack hole: diameter 40 miles

Large moon crater: diameter 120 miles

Largest asteroid: diameter 620 miles

...ars: diameter 4,217 miles

White dwarf: diameter 5,000 miles

Venus: diameter 7,521 miles

COMETS, ASTEROIDS, AND METEORITES

This volume is one of a series that examines the universe in all its aspects, from its beginnings in the Big Bang to the promise of space exploration.

VOYAGE THROUGH THE UNIVERSE

COMETS, ASTEROIDS, AND METEORITES

BY THE EDITORS OF TIME-LIFE BOOKS
ALEXANDRIA, VIRGINIA

CONTENTS

1/CELESTIAL VISITORS

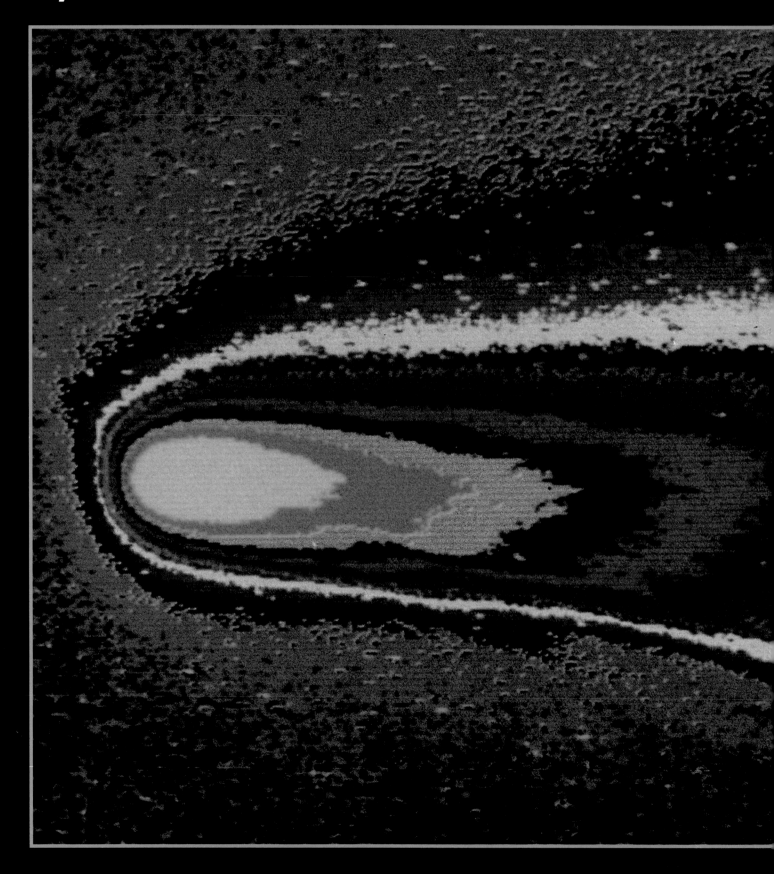

Halley's comet blazes with computerized colors in this photograph taken from a ground-based telescope on March 4, 1986, three weeks after the comet's closest approach to the Sun. The various hues, which indicate decreasing brightness outward from the head, help reveal details in the structure of the tail.

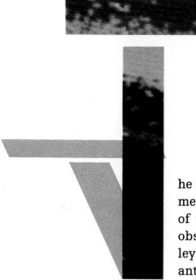

he return to Earth of Halley's comet in 1986 was, by some measures, a monumental fizzle. Primed by media accounts of the comet's previous visit in 1910, when earthbound observers had been treated to spectacular views of Halley's head and 30-million-mile-long tail, the public was anticipating an impressive light show. But when the comet finally hove into view on November 8, 1985, it presented itself as a mere smudge of light alongside the Pleiades star cluster. Most people never even caught so much as a glimpse of the Solar System's most famous periodic guest, which looped around the Sun the following February, then skidded within 39 million miles of Earth on April 10 and was gone, headed back toward the distant planets and its next scheduled terrestrial rendezvous in 2061.

For the international science community, however, Edmond Halley's comet staged a magnificent show. As the comet reached perihelion (its point of closest approach to the Sun) and turned to head past Earth, exhilarated astronomers made their own odyssey, crossing international boundaries to travel from one control center to another to witness closeup images beamed back by an argosy of reconnaissance space probes. These craft—the Soviet Vega 1 and Vega 2, the Japanese Suisei and Sakigake, and the European Space Agency's Giotto—used a wide range of sophisticated instruments, from cameras and spectrometers to particle detectors and magnetometers, to capture the celestial vagrant in glorious detail.

Seen in computer-enhanced color on television monitors at the Soviet Space Research Institute in March, shortly after perihelion, and a few days later at the European Space Agency monitoring center in Darmstadt, West Germany, Halley's comet was stripped of the mystery of the ages. Here loomed a misshapen chunk of matter, ten miles long and five miles thick, that resembled a peanut, a down-at-the-heels old shoe, or a legless water buffalo, depending on which astronomer happened to be describing it. The comet's nucleus, marked by a crater 600 feet deep and a hill even higher than that, wobbled about its axis once every few days, shedding dust and spewing out gases that enrobed it in a spherical cloud called a coma. As with every comet, this exhaust also streamed away behind the nucleus to form a pair of tails: a straight gas tail that can reach 50 million miles in length and a second, stubbier, curved tail made up of particles of dust.

Some theorists had imagined that the comet's nucleus would be pale and highly reflective because of the ice it contained, but the object they saw was instead as black as coal—darker, in fact, than almost everything else in the Solar System. Of the sunlight striking the nucleus, just three percent was reflected. This unexpectedly low reflectivity would prove to be just one of many surprises that Halley had in store.

TO SEEK OUT LESSER WORLDS

Since the early 1960s, when space probes first opened their mechanical eyes on the Solar System, astronomers have witnessed all manner of violent and inspiring spectacles, from the Sun's explosive flares to swirling storms on Jupiter. Though dramatic, the escapades of these giants merit only slightly more attention than do those of the Solar System's three classes of smaller inhabitants: the comets and asteroids that swarm the interplanetary expanses and the meteoroids that streak into Earth's atmosphere from space. (Meteoroids that become visible as they penetrate Earth's atmosphere are called meteors; those that survive their plunge to the ground are known as meteorites.) Scientists calculate that the three classes, likely relics of the Solar System's creation, exist in phenomenal abundance. The total number of comets alone may be as high as 10 trillion, and their combined mass may be seven times that of Earth.

Comets have been objects of study since antiquity, but astronomers did not discover the first asteroid until 1801. After two centuries of observation, they have identified nearly 15,000 of these small, rocky bodies, ranging in diameter from less than 1 to about 600 miles. However, so many of the smaller ones have eluded detection that the total number may actually be in the millions. Whatever the exact figure, scientists have determined that the asteroids—which orbit the Sun in belts between Mars and Jupiter—have a combined mass that is approximately one-twentieth that of the Moon.

Meteoroids, the third contingent of the lesser-worlds clan, derive from both comets and asteroids. They may be sand-size particles of cometary debris that have continued to navigate the Solar System for eons after leaving their parent body, or they may be larger fragments that resulted from collisions between asteroids traveling in perilously close orbital paths. In the twentieth century, such stones from space would spawn a new branch of astronomy dedicated to understanding the parent bodies of meteorites through the study of their recovered offspring.

The earliest known record of a comet sighting was made by a Chinese court astrologer in 1059 BC. The Chinese called comets "broom stars" not only for their gently flared profile but also because comets were thought to herald the sweeping away of the old and the arrival of the new. For the next two millennia, the Chinese faithfully cataloged the movements of comets; by the eleventh century AD, court astrologers—who were also, in effect, astronomers—had produced a sizable compendium of sightings. This and other records kept by the Chinese described 600 separate observations and established the fact that comet tails always point away from the Sun.

The ancient Greeks considered comets to be harbingers of dire events. A Homeric couplet written around 900 BC, for example, compared the helmet of the warrior Achilles to "the red star that from [its] flaming hair / Shakes down disease, pestilence and war." Later philosophers adopted a more scientific—though no less flawed—outlook. In the fourth century BC, for instance, Aristotle posited that comets were a form of weather, an atmospheric disturbance caused when earthly emanations were set ablaze in the region of fire that he believed surrounded the planet.

Of the many observers who challenged this interpretation, Seneca—a tutor and political adviser of the Roman emperor Nero—offered the most cogent critique. Cometary movements resemble those of the planets, Seneca argued in the first century AD, so comets cannot possibly be atmospheric in nature; instead, they must occupy the distant realms of the cosmos.

Despite the weight of Seneca's analysis, the Aristotelian view held sway until 1577, when a Danish nobleman named Tycho Brahe performed a thorough study of a particularly bright comet that lit the skies over northern Europe for three months. As he charted the comet's path, Brahe realized that a comet plying a course within Earth's atmosphere would appear in markedly different positions against the background stars when viewed by observers at different places on the globe. This illusion, known as parallax, operates in daily life: An object held close to the face seems to shift across one's field of view when it is viewed first with one eye alone and then with the other; as the object is held farther and farther away, however, the shift grows less and less pronounced until it vanishes altogether. By correlating his own measurements of the comet's progress across the sky with reports from colleagues

A cometary catalog, painted on silk by Chinese astronomers in 168 BC and discovered in a tomb near Changsha in 1973, depicts twenty-nine different types of comets (eight of which are shown here) and lists their associated omens. Four-tailed varieties portend "disease in the world," while those with three tails bode "calamity in the state." A comet that has two tails curving out to the right *(far left)* foretells minor warfare—but "the corn will be plentiful." The Chinese began keeping records of comets as early as 1059 BC.

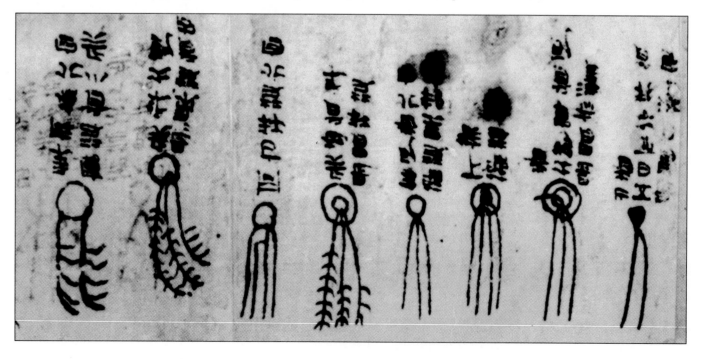

ISTI MIRANT STELLÃ

HAROLD

A section of the Bayeux tapestry, which commemorated the Norman conquest of England in 1066, portrays the appearance of a comet—now known to be Halley's—shortly before the invasion began. While his people stand in awe of the cosmic visitor (the Latin inscription reads, ''They marvel at the star''), King Harold of England receives word of the fateful omen. For Harold, the comet was indeed a harbinger of doom: He was killed that year in the Battle of Hastings.

around Europe, Brahe proved that this comet was subject to very little parallax. He therefore concluded that it must be traveling at great distances from Earth—at least four times as far away as the Moon.

MR. HALLEY PAYS A VISIT

Tycho Brahe's estimate of one million miles for the distance of one comet was grossly short of the actual ranges of comets in general, but his refutation of Aristotle's outlook was essentially correct. It was also inspirational, providing intellectual grist for a group of English astronomers, physicists, and chemists who had begun holding weekly meetings in London during the mid-1600s to discuss issues of common interest. In 1660 they were granted a royal charter to form the Royal Society for the Improvement of Natural Knowledge. They began debating, among other topics, the question of cometary motions.

No one showed greater interest in the subject than young Edmond Halley (rhymes with ''valley''), the son of a well-to-do soap boiler and meat salter who had made his fortune provisioning the British navy. Halley displayed such a facility with matters of science that when he reached the age of seventeen his father presented him with a twenty-four-foot-long telescope

Among the most spectacular comets of the twentieth century, Ikeya-Seki emblazons the early-morning sky above the town of Flagstaff, Arizona, on October 28, 1965. (City lights appear at the bottom of the image.) Its dust tail stretched for millions of miles and remained visible for several months, bright enough to be seen well into the daylight hours. Ikeya-Seki did not survive its solar encounter intact; strained by the Sun's intense gravitational force, its nucleus split in two shortly after perihelion.

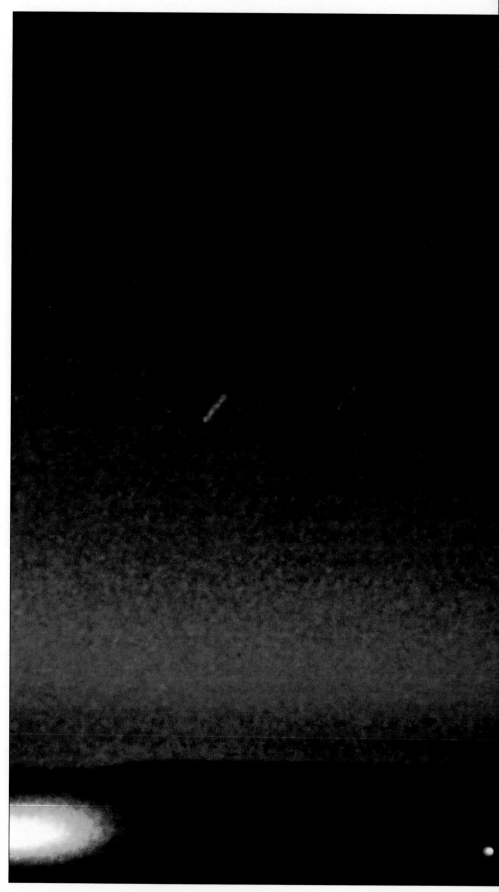

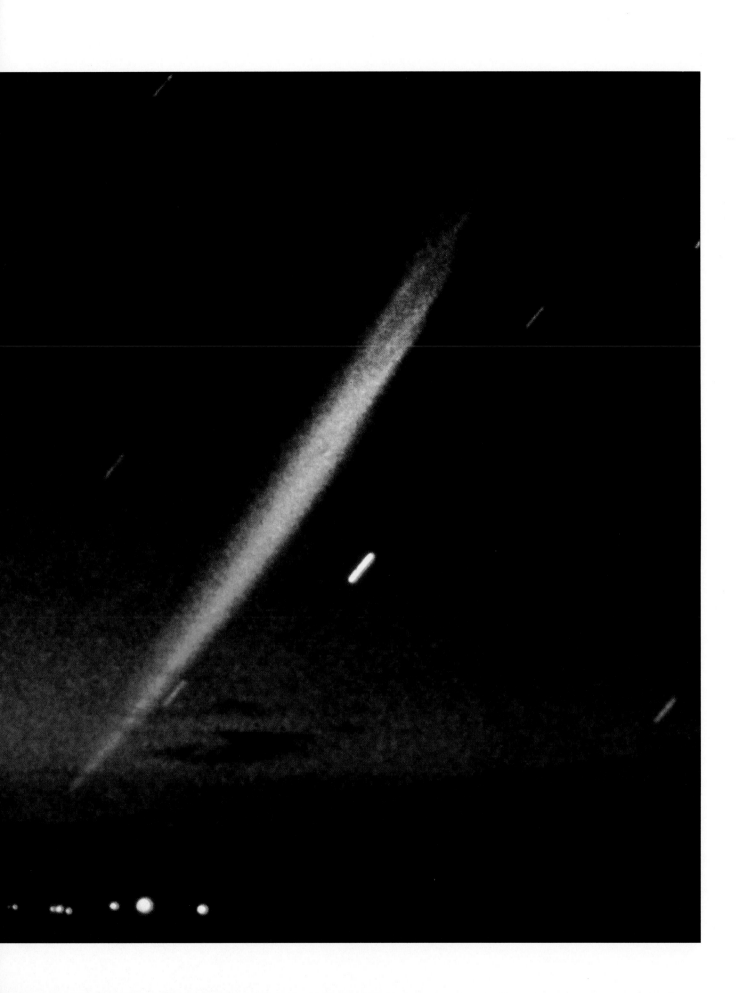

and sent him off to Oxford University. Two years later, in 1675, Halley announced to the Astronomer Royal, John Flamsteed, that he had discovered errors in the published positions of the planets Jupiter and Saturn as well as in star charts compiled by Tycho Brahe. Based on this early work, and on Halley's publication in 1678 of the first star chart prepared with the aid of a telescope, the Royal Society judged Halley—then just twenty-two years old—worthy of admission to the elite group.

In the 1680s, Halley and fellow society members Christopher Wren (later the architect of Saint Paul's Cathedral) and physicist Robert Hooke began trying to puzzle out the fine points of planetary motions, whose broad mathematical outlines had been set forth in the early 1600s by German astronomer Johannes Kepler. The trio suspected that the Sun's gravity operates on planets—and, by extension, on comets—in a fashion similar to that by which light is propagated, weakening proportionally over distance. Moving a given planet twice as far from the Sun would thus reduce the pull of solar gravity by a factor of four, they surmised, while moving the planet three times as far would reduce the force by a factor of nine. This supposition—which has since been formalized as the inverse-square law—made intuitive sense, yet the three men could not prove it mathematically. Stymied, Wren in 1684 offered a prize—a book of the winner's choice, as long as it cost no more than forty shillings—to anyone who could do so.

Halley took a less mercantile approach: He traveled to Cambridge University to seek the assistance of Isaac Newton, whose genius was rivaled only by his reclusiveness. Newton assured Halley that gravity could indeed be described by the inverse-square law. When Halley asked the mathematician how he could be so certain, Newton replied that he had worked out the proof seventeen years earlier but had misplaced it among his papers. At Halley's urging, however, Newton was able to reconstruct his calculations. For the next three years, Halley acted as Newton's agent, confidant, and financier to ensure the publication of Newton's masterwork, *Philosophiae Naturalis Principia Mathematica*, meaning "the mathematical principles of natural philosophy." In *Principia*, Newton built on Kepler's laws of planetary motion to demonstrate that the motions of comets could be described as following parabolic paths with the Sun at one focus. From this, both Newton and Halley inferred that all comets follow highly elliptical orbits.

At about this time, Halley began examining orbital data for a handful of comets. Because the calculations necessary to compute cometary orbits were tedious and time-consuming, he approximated the comets' true elliptical orbits by computing their parabolic trajectory around the Sun; this required five rather than six variables to calculate.

His investigations turned up a particularly intriguing coincidence. He noticed that the comet of 1682—whose route and date of perihelion had been carefully established by European astronomers—followed a path almost identical to the one taken by two other comets, which had appeared in 1607 and 1531. In 1695, Halley wrote Newton of his strong suspicion that the three were in fact one: They occurred roughly seventy-five years apart, and each described the same retrograde orbit—that is, one contrary to Earth's passage around the Sun.

Although Halley made a formal presentation of these findings to the Royal Society in 1696, he was unable to devote his full attention to comets until eight years later, when he was appointed to the chair of geometry at Oxford University. At that time, he threw himself into his studies, examining orbital data on twenty-four comets that had been observed from 1337 to 1698.

In 1705, he published "A Synopsis of the Astronomy of Comets," a pamphlet in which, among other things, he accounted for the failure of his seventy-five-year comet to return at rigorously precise intervals. The gravitational influence of the giant planets Jupiter and Saturn, he reasoned, would slightly perturb, or distort, the comet's flight, advancing or retarding each successive Earth visit by a matter of several months. He predicted that the comet would next be seen in late 1758 or early 1759.

By the time of Halley's death in 1742, his forecast had taken on a life of its own, and professional astronomers throughout Europe were eagerly antici-

COMET TRAILBLAZERS

A host of great thinkers, from philosophers and inventors to mathematicians and physicists, have helped lift the veil of mystery surrounding the Solar System's icy wanderers. After many centuries of evoking primarily fear and superstition, comets began to attract serious scientific attention in the late 1600s. For early investigators, the subject stirred little more than a passing interest; Edmond Halley, for example, devoted much of his time to other pursuits, such as designing and testing deep-sea diving bells. Significant milestones were relatively sporadic over the next 250 years, but the 1950s ushered in a golden age of cometary theorizing. Building on the achievements of their predecessors, a few key figures worked out the answers to the most basic questions about the origin, composition, and behavior of comets.

1705 English scientist Edmond Halley predicted the return fifty-three years thence of the comet that now bears his name. He died in 1742, sixteen years before he would be proved right.

1755 Best known for his works on philosophy, the German intellectual Immanuel Kant was the first to postulate that the Solar System condensed from a cloud of interstellar matter, whose outermost regions spawned the comets.

pating the comet's return. True to Halley's estimate, the comet reappeared in 1758, first spotted by a German farmer through a homemade telescope on Christmas night. Halley's name was attached to the visitor, fulfilling a second prediction of his: that "candid posterity" would "not refuse to acknowledge that [the first periodic comet] was discovered by an Englishman."

WHENCE COMETS?

Although a comet's orbit could now be plotted with relative certainty, its origins remained mysterious: Are comets born inside the Solar System? Or does the cometary nursery lie beyond, in the expanses of interstellar space? These questions, touching on the creation of the Solar System itself, would not be convincingly answered until the twentieth century.

Many eighteenth-century scientists subscribed to the cosmogonal view of French polymath René Descartes, who in 1644 had suggested that the Sun and its clutch of planets coalesced from a spinning cloud, or nebula, of gas and dust. Comets, Descartes believed, had formed from dead stars and were denser than the planets.

The Cartesian view attracted a number of theoretical fine-tuners, among them Immanuel Kant. Kant is generally remembered today for the philosoph-

1863 A specialist in celestial mechanics, American Hubert Newton refined Laplace's ideas about Jupiter's influence, showing that very few distant comets would be so affected. A paper by Newton helped establish the link between meteor showers and comet orbits.

1813 French mathematician Pierre-Simon de Laplace proposed that far-flung comets are swept into the inner Solar System by Jupiter's gravity.

1836 Explaining why comets do not always return when predicted, German astronomer Friedrich Bessel noted that they may release jets of material that alter their orbits.

ical treatises he penned later in life, but as a young man he conducted some profoundly original explorations in the field of astronomy. Using Descartes's scheme as his point of departure, Kant imagined that the Solar System had condensed from a vast cloud of what he called "primitive matter." He maintained that comets were spawned not from dead stars but from the diffuse material at the periphery of this cloud, where their great distances from the strong tug of gravity at the cloud's center would allow them to fly through space in "lawless freedom."

A second elaborator of Descartes's nebular theme was Pierre-Simon de Laplace, a French scientist who was versed in math, astronomy, and physics. In 1796, Laplace postulated that the rotating solar nebula had accelerated as it contracted, throwing off various rings of gas and dust that later accreted to form the planets and their moons. Laplace did not include comets in his scheme until 1813. After studying changes that had occurred in the orbit of comet Lexell when the body passed near Jupiter in 1767 and again in 1779, he inferred that the tug of the giant planet might have deflected many comets from their original, more distant paths into shorter orbits within the confines of the Solar System.

Laplace's work set the stage for the current distinction between long- and short-period comets. Long-period comets venture into the inner Solar System from aphelia, or orbital extremes, that extend up to five trillion miles from the Sun. Short-period comets—especially those with orbits of less than thirty years—behave much like the planets. A large proportion have aphelia near

1870s Based on evidence that comets breed meteor showers, British astronomer Richard Proctor *(far left)* asserted that comets are loose clusters of meteoroids. His countryman Raymond Lyttleton was pushing the now-discredited gravel-bank theory eighty years later.

1880s Russian astronomer Fedor Bredikhin developed the first modern classification scheme for comet tails, which distinguished them by their length and their degree of curvature.

Jupiter and move in a prograde fashion; that is, they circle the Sun in the same direction as do the major planets. (Halley's comet, with its retrograde movement, is one of the five known exceptions.) The comet with the shortest period is Encke. Its perihelion lies inside the orbit of Mercury, and its aphelion falls at a point between Mars and Jupiter, allowing it to race through one full orbit in just 3.3 years.

OF JETS AND TAILS

Astronomy in general and cometary science in particular experienced a boom in the nineteenth century with the advent of two new technologies. The techniques and instruments of spectroscopy enabled scientists to determine the elemental composition of an object by analyzing the light it emitted or absorbed, and the development of photography permitted them to record on film celestial phenomena invisible to the naked eye. Aided by these tools, researchers painstakingly assembled a more detailed picture of the structure and chemical makeup of individual comets.

First, they had to explain the disturbing departure of some comets from the clockwork motions described by Isaac Newton's fundamental laws of physics. In the early 1800s, German astronomer Johann Franz Encke set out to calculate the orbit of the comet that now bears his name. Encke carefully factored in the complex gravitational pulls that the comet would experience as it

1908 British physicist Arthur Stanley Eddington discovered clumps of material moving down the tail of comet Morehouse at unexpectedly high speeds, indicating that a force more powerful than the pressure of sunlight drives comet tails away from the Sun.

1950 Dutch astronomer Jan Oort proposed that many comets derive from a vast spherical cloud of comets enveloping the Solar System.

1950 Fred Whipple, the American guru of comet science, introduced the idea that comet nuclei are made up of ice and dust particles, relatively unaltered chunks of the Solar System's raw materials. This model, called the dirty-snowball, still remains the accepted description.

darted among the inner planets, yet his timetable proved to be flawed: With every return, the comet reached its perihelion two and a half hours later than Encke's math said it should. At a loss for an explanation, Encke proposed the existence of an undetected "resisting medium" that was slowing the comet during its perihelion passages.

The reappearance of Halley's comet in 1835 prompted German mathematician Friedrich Wilhelm Bessel to advance an alternative explanation. As the comet neared perihelion, Bessel noticed, jets of material spurted from its head. Bessel wondered whether these jets might act as rockets, forcing comets into smaller orbits or boosting them into larger ones, depending on the time and direction in which the jets fired. His notion was essentially correct; so circuitous is the course of scientific advance, however, that the theory would lie dormant for more than a century.

Bessel also noted that as the jets shot sunward from the comet's head, the material they contained was invariably swept back in the opposite direction, as if it was being repelled by some force emanating from the Sun. He then calculated the parabolas that should result from such interactions between the Sun and a comet. Bessel's conjecture was borne out by the appearance of comet Donati in 1858: For several weeks, the comet's head was veiled by fountains of material that conformed almost precisely to Bessel's theoretical plottings.

Bessel's studies inspired generations of scientists to pursue their own investigations of cometary structure. In about 1885, for example, Russian astronomer Fedor Bredikhin divided comet tails into three broad classes. Type I tails, he observed, streamed out in a nearly straight line, changed shape from day to day, and exhibited bulges, knots, and kinks. They could grow enormously long, as had the tail of the Great March comet of 1843, which ran

1951 German astrophysicist Ludwig Biermann posited the existence of the solar wind, a stream of charged particles from the Sun that is capable of giving rise to a comet's plasma tail.

1957 Swedish plasma physicist Hannes Alfvén elaborated on Biermann's theory, explaining that magnetic fields carried by the solar wind sculpt comets' tails of ionized gas.

from the horizon to the zenith—about 180 million miles in absolute terms—when seen from the latitude of Bombay, India. Type II and III tails, said Bredikhin, were highly curved, stable, and relatively short, no longer than about six million miles. Modern astronomers have combined the last two types under a single rubric—dust tails—while Bredikhin's Type I tails are now known to be made up of gas.

Bredikhin hazarded (erroneously, as it turned out) that comet tails of all three types consist of gas molecules. He also guessed—correctly, this time—that some unspecified force emanating from the Sun repelled the molecules differentially according to their weight: The light molecules of the straight tails were dispersed faster and farther than the heavier molecules that made up the curved tails.

A clue to Bredikhin's mysterious solar force was supplied by Swedish chemist Svante Arrhenius some years later, in 1900. Building on earlier work by Kepler and Scottish physicist James Clerk Maxwell, Arrhenius proposed that sunlight exerts enough mechanical pressure on the microscopic particles in a comet's tail to keep the tail constantly pointing away from the Sun. His view seemed to be corroborated by a group of laboratory experiments conducted in 1901, which confirmed that light pressure can indeed influence the motion of such particles.

Just seven years later, however, the near-Earth passage of comet Morehouse put Arrhenius's theory to a tougher test. Among the luminaries who gathered at Oxford University to discuss the comet's passage was Arthur Stanley Eddington, soon to emerge as Britain's preeminent physicist and the twentieth century's most articulate champion of Einstein's theory of relativity. After scrutinizing photographs of the highly active comet, Eddington noted that its nucleus was periodically rocked by convulsions or explosions of some sort, in the aftermath of which large bulges of material passed rapidly down its tail. When Eddington calculated the velocities involved, he realized that the material must be propelled by a force stronger than mere light pressure—although what, he could not say.

A GRAVEL BANK IS BORN

Concurrent with the nineteenth-century quest to puzzle out the overall structure of comets there emerged the first widely accepted theory of cometary composition. The theory hinged on observations made during the 1860s, when scientists first linked comets to meteor showers. In 1863, a Yale College professor named Hubert Newton noticed that the Leonid meteor showers, which ordinarily rain about a dozen meteors per hour into Earth's atmosphere around November 17 each year, peaked in intensity every thirty-three years, when as many as 10,000 meteors could occur in an hour. To Newton, the pattern suggested that Earth was passing through the densest part of a cloud of orbiting debris. Then in 1866, Italian astronomer Giovanni Schiaparelli showed that the meteors of another annual shower—the Perseids, which swarm in each August from the direction of the constellation Perseus—were

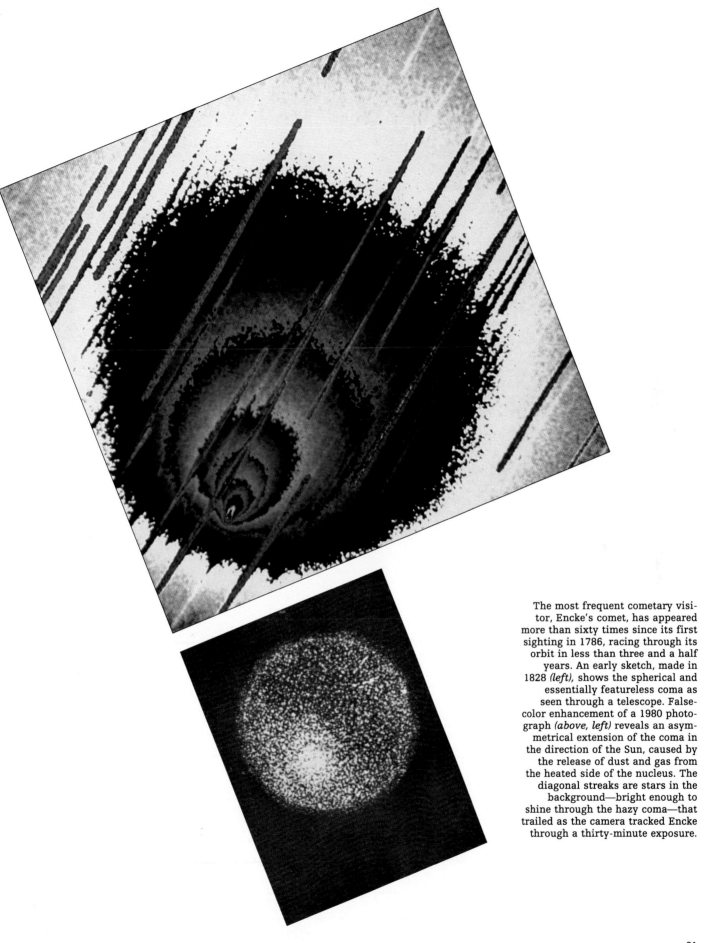

The most frequent cometary visitor, Encke's comet, has appeared more than sixty times since its first sighting in 1786, racing through its orbit in less than three and a half years. An early sketch, made in 1828 *(left),* shows the spherical and essentially featureless coma as seen through a telescope. False-color enhancement of a 1980 photograph *(above, left)* reveals an asymmetrical extension of the coma in the direction of the Sun, caused by the release of dust and gas from the heated side of the nucleus. The diagonal streaks are stars in the background—bright enough to shine through the hazy coma—that trailed as the camera tracked Encke through a thirty-minute exposure.

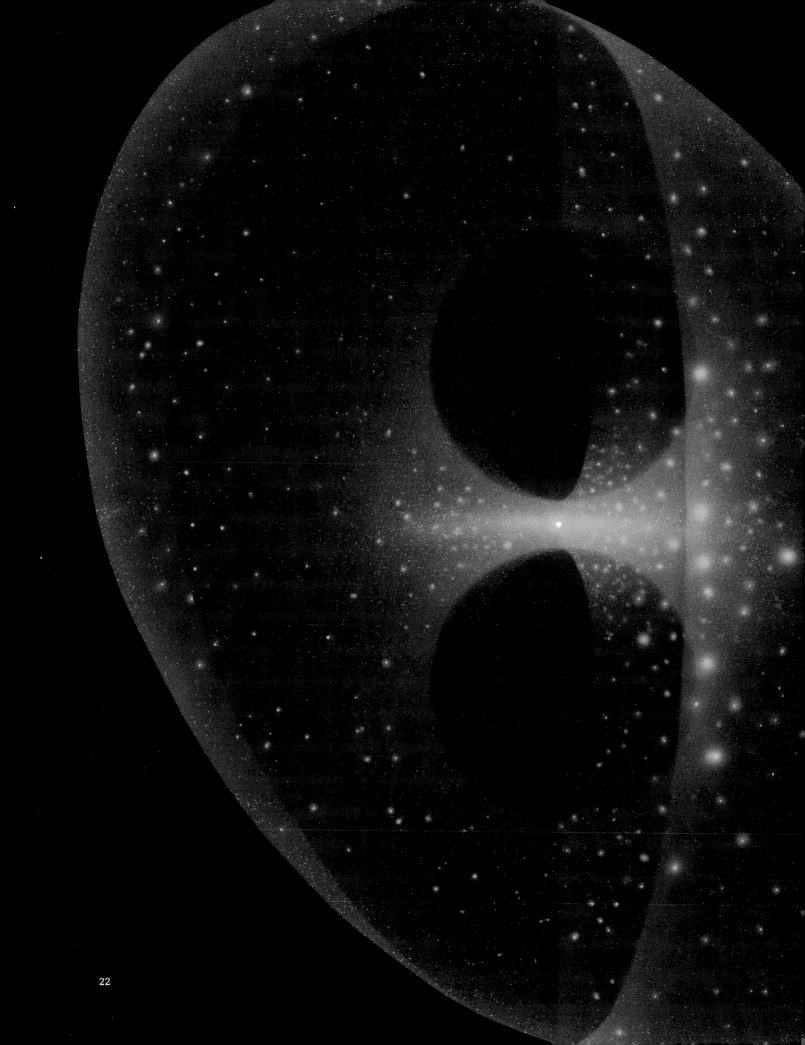

Genesis of a Cloud of Comets

the nearly four centuries since the advent of the escope, about 150 comets have visited the inner lar System more than once. Another 600 have made single appearance during that time. But these so-led short- and long-period comets represent only a nuscule fraction of the total comet complement. Scientists believe that the comet population numbers in the trillions, enveloping the Solar System in a vast swarm *(left)* whose outermost boundaries reach halfway to the nearest star.

Recently, with the help of powerful computers, astronomers have modeled the orbital dynamics of the three regions they hypothesize within the overall cometary system *(pages 24-27)*. Just beyond the orbit of Uranus lies the billion-mile-

Gerard Kuiper, who first suggested that debris left over from the formation of the Sun and planets would be found there. Beyond Neptune, the Kuiper belt flares to form the so-called inner cloud, which may extend outward as far as 20,000 astronomical units (20,000 times the distance between the Sun and Earth). The inner cloud in turn merges with the Oort cloud, hypothesized by Dutch astronomer Jan Oort in 1950, which reaches 50,000 AU into space—so far that comets are held to the Sun by the flimsiest of gravitational tethers.

Gravitational forces within the Solar System, as well as those resulting from its location and orientation in the disk of the Milky Way galaxy *(below)*, determine how large and how elliptical a comet's orbit is and how much it is inclined to the plane of the Solar System. These factors in turn determine how frequently a particular comet will show itself in terrestrial skies—or whether someday it might be stripped from the solar family altogether.

GRAVITATIONAL EXPULSION

When Jan Oort postulated the existence of the Oort cloud, he and other astronomers assumed that it was the repository for all periodic comets that visit the inner regions of the Solar System. Recently, however, computer simulations by a three-person team from Lick Observatory and the University of Toronto have demonstrated that many short-period comets probably reside in the Kuiper belt, the presumptive birthplace of all comets.

During the early life of the Solar System, the region roughly thirty astronomical units from the young Sun was distant enough for gaseous remnants of the solar nebula to have frozen into cometary nuclei. Over the tions illustrated here and on the following three pages, most of the denizens of this flattened, circular belt were expelled. Some were flung out of the Solar System to drift among the stars, but trillions of them adopted the enormously elongated orbits that now sketch the dimensions of the Oort cloud. Of the comets that never left the Kuiper belt, a few were pulled by gravitational interaction with the large planets in the outer Solar System into smaller, more elliptical solar circuits that brought them past Earth every few score years or oftener. The diagrams at right illustrate how the giant planets can influence the orbits of Kuiper belt comets, either shrinking the orbits to pull the comets in toward the Sun or expanding them so that the comets leave the Kuiper belt and become susceptible to other objects and forces in the galaxy *(pages 26-27)*.

100 AU

As a comet orbiting the Sun in the Kuiper belt passes in front of Uranus, its forward motion is retarded by the planet's gravitational pull. With the resulting loss of energy, the comet's orbit shrinks until its perihelion, the point in its orbit nearest the Sun, lies inside the orbit of Saturn. Saturn's gravitational influence in turn hands the comet off to Jupiter, completing the multistep process represented by the dotted line and shrinking the comet's orbit until its aphelion lies near Jupiter's orbit.

100 AU

When a comet passes behind a large planet (in this case, Neptune), the planet's gravitational force can act to enlarge the comet's orbit *(dotted line)*. Since the planet's pull is in the same direction as the comet's forward motion, the comet gets an energy boost sufficient to extend its aphelion to as much as several thousand astronomical units. With loosened ties to the Sun, the comet is now prey to other gravitational influences.

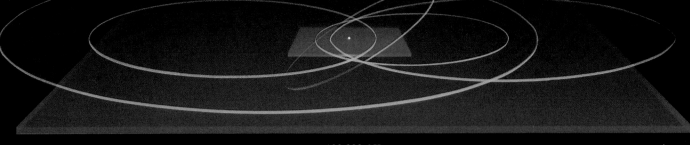

100,000 AU

Starting at about 1,000 AU from the Sun—more than thirty times farther out than the orbit of Neptune—passing stars and giant molecular clouds affect the size of comet orbits over time. But the strongest influence on orbital expansion is believed to come from gravitational interactions with the mass of the galactic disk *(below)*. These three forces also contribute to increasing the inclination of comet orbits to the ecliptic plane. Comets that originated in the Kuiper belt now supply the inner cloud *(right)*. Similar interactions then eject some of these comets into the more distant Oort cloud *(right, bottom)*.

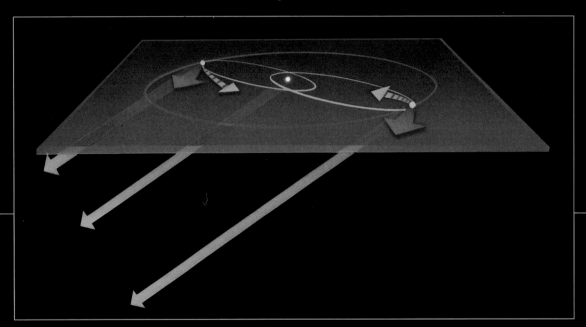

THE GALAXY'S TIDAL PULL

The massive bulk of the Milky Way galaxy's central plane lies below the plane of the Solar System, tilted at an angle of 60 degrees, thereby exerting differential gravitational effects known as tidal forces *(green arrows)* on the Sun and its halo of comets in the Oort cloud *(purple circle)*. As illustrated here, comets are most susceptible to galactic tidal forces when they are at aphelion—their farthest point from the Sun.

A comet with aphelion in the part of the Oort cloud nearer to the galactic disk *(above, left)* feels a component of the tidal force *(orange arrow)* that effectively tugs in the same direction as the comet's orbital motion, a boost that kicks it into a larger orbit *(blue dashed arrow)*. For a comet on the opposite side of the Oort cloud, in contrast *(above, right)*, that component of the galaxy's tidal force acts against the comet's direction of motion, decreasing its energy and shrinking its orbit—until it dips into the inner regions of the Solar System as a long-period comet that may take millions of years to complete one circuit.

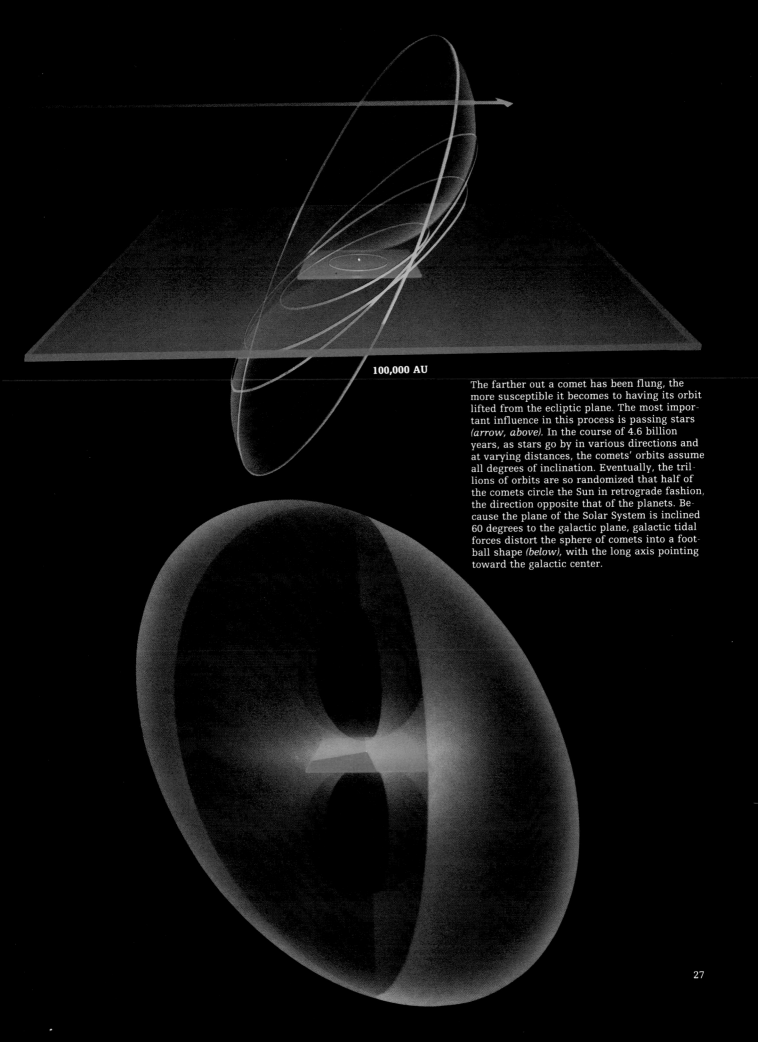

100,000 AU

The farther out a comet has been flung, the more susceptible it becomes to having its orbit lifted from the ecliptic plane. The most important influence in this process is passing stars *(arrow, above)*. In the course of 4.6 billion years, as stars go by in various directions and at varying distances, the comets' orbits assume all degrees of inclination. Eventually, the trillions of orbits are so randomized that half of the comets circle the Sun in retrograde fashion, the direction opposite that of the planets. Because the plane of the Solar System is inclined 60 degrees to the galactic plane, galactic tidal forces distort the sphere of comets into a football shape *(below)*, with the long axis pointing toward the galactic center.

27

moving in orbits similar to that of comet Swift-Tuttle. The next year, Schiaparelli tied the occurrence of the Leonid meteor showers to the orbit of comet Tempel-Tuttle, which rounds the Sun every thirty-three years. Schiaparelli had discerned that meteor showers were the result of Earth's passage through the so-called exhaust trail of cometary debris that tends to spread throughout a comet's orbit. Because the debris is densest in the area just behind the comet, the meteor showers are most intense just after a comet's periodic appearance in terrestrial skies, when Earth crosses this section of the comet's path.

Around the end of the decade, English astronomer Richard Proctor and others drew on the newly established connection between meteor showers and comets to make two bold assertions: First, Proctor correctly proposed, the material that causes a meteor shower must follow an orbit similar to that of a comet; second, he suggested with considerably less accuracy, this material is the very stuff of comets—and therefore a comet is nothing more than a clump of tiny meteoroids.

Proctor's "flying gravel-bank theory," as it became known, found a champion in British theoretician Raymond Lyttleton. According to the Proctor-Lyttleton doctrine, a comet is made up of small, solid particles that fly in formation with one another while orbiting the Sun. Lyttleton could not explain, however, the jets of material that Bessel and others had detected shooting from the heads of comets. Even if the gravel chunks had absorbed gases in deep space, they could not possibly retain the vapors during multiple passes through the superheated precincts around the Sun. Yet Halley's comet, for one, put on brilliant shows century after century.

Lyttleton also had difficulty postulating a mechanism whereby a gravel-bank comet could change its orbit in the fashion of comet Encke. A final failing of gravel-bank anatomy was that it could not accommodate the documented existence of "sun grazers"—comets that literally skim the Sun's surface at their perihelion points. Surely the unfathomable heat of such an encounter would obliterate a body made up of small particles?

SHAPING THE DIRTY SNOWBALL

Despite these logical gaps in the gravel-bank concept, the lack of a better alternative enabled it to survive until 1950. In that year and the next, however, a trio of theoretical breakthroughs began to redefine the entire field of cometary science.

The first was Jan Oort's assertion that the Solar System is enveloped by an almost unimaginably distant—and inconceivably populous—swarm of comets. Oort, an eminent Dutch astronomer who had deduced both the spiral structure and the rotation rate of the Milky Way galaxy, posited in 1950 that a spherical cloud containing billions upon billions of comets existed at a distance of what was then calculated to be 50,000 to 150,000 astronomical units from the Sun. (An AU equals the average distance from the Sun to Earth, or about 93 million miles.) These unseen multitudes, Oort proposed, were gravitationally anchored to the Sun, but they followed such immense orbits—some reaching almost halfway to the nearest star, Alpha Centauri—that they required millions of years to complete a single circuit.

The comets had probably started life somewhere in the asteroid belt, Oort's theory went, only to be expelled by Jupiter's gravity into that distant cloud; there they would reside until the gravitational attraction of a passing star happened to disturb their orbits and so dislodge them. Comets ejected from the Oort cloud in this manner would then embark on 10-million-year-long treks toward the Sun; as the comets neared the inner Solar System, the pull of the larger planets—Saturn and Jupiter in particular—would snare a number of them into long- or short-period orbits.

Oort had even attempted a rough census of the number of comets that would populate his theorized cloud. In any given period of one million years, he

COMET DONATI'S MULTIPLE HALOS

Like water cascading from a fountain, luminous streams of dust and gas flow back from the head of comet Donati in these drawings made in 1858 by observers in Russia *(left)* and at the Harvard Observatory *(below, left and right)*. Scientists speculate that as the nucleus spins, it periodically exposes a volatile region on its surface to sunlight, unleashing a fresh batch of ejected material with every rotation. The pair of sketches below, representing the comet on two successive days, shows a series of concentric halos that formed 4.6 hours apart—the presumed rotation rate of the nucleus.

calculated, about five to ten neighboring stars would pass close enough to the Oort cloud to perturb some of its comets into the inner Solar System. He combined this figure with his estimate of the number of new long-period comets—about one per year—that were observed to approach the Sun. In order to maintain that flux, Oort's calculations showed, the comet cloud would have to be some 190 billion strong in membership. The number of comets in the Oort cloud is thus roughly equivalent to the total number of stars in the Milky Way galaxy.

In 1951, astronomer Gerard Kuiper of the Yerkes Observatory in Wisconsin suggested a cometary fount much closer to home than the Oort cloud. Kuiper accepted the recently developed theory that the Sun had gone through a period of particularly intense activity early in its formation; during this T Tauri phase, as it is called, the newborn Sun burned with ten times its current brilliance for tens of millions of years. Kuiper speculated that the Sun's increased heat would have vaporized all of the volatile primordial gases that once existed in the inner Solar System, depriving potential comets of the congealed gases necessary to form a nucleus. He then postulated that comets originated instead in a flattened disk of solar-nebula debris just beyond the orbit of Neptune, about thirty-five to fifty astronomical units from the Sun, where low temperatures would have encouraged primordial gases to condense into cometary nuclei. Evidence supporting Kuiper's revolutionary idea would be delivered by a complex computer model nearly four decades later.

In the meantime, precisely what those comet nuclei might consist of was outlined in a second breakthrough, this one by astronomer Fred Whipple, a former meteor specialist who would ultimately emerge as the dean of cometary scientists. From 1949 to 1956, Whipple chaired the Harvard University astronomy department; later he served as director of the Smithsonian Institution's Astrophysical Observatory in Cambridge, Massachusetts. In 1950, Whipple unveiled his trailblazing theory, dubbed the "dirty-snowball model," which held that a comet's nucleus was an agglutination of frozen water, ammonia, methane, carbon dioxide, and hydrogen cyanide, a leftover lump of the protean Solar System, manufactured in the deep freeze of space. With its icy composition making it a poor conductor of heat, Whipple theorized, the nucleus warmed unevenly as it approached the Sun, gradually evaporating to yield the blowzy coma and the spurting gas jets. According to Whipple's model, the vapors issuing from a comet's heated nucleus extruded dust particles that formed its tails, which would tatter and diminish as the comet headed back toward aphelion; at that orbital extreme, temperatures would be too low to maintain vaporization and the tails would disappear.

Whipple's model neatly explained how the interaction between a comet's nucleus and the Sun created the coma, jets, and tails, and how a comet could survive repeated visits to that star. During most of its orbit, said Whipple, a comet's constituent matter was conserved; only during its brief pass-

Stirred to activity by the Sun's warmth, comet Swift-Tuttle releases jets of fine-grained dust as indicated by these sketches from its 1862 manifestation. The outbursts apparently emanate from several small, isolated areas on the nucleus. Dust shed by the comet lingers in the inner Solar System and is responsible for the Perseid meteor shower each August, when Earth crosses the comet's orbital path. This connection was first noted in 1866 by Italian astronomer Giovanni Schiaparelli, who also linked November's Leonid meteors to comet Tempel-Tuttle.

es by the Sun did the nucleus slough off massive amounts of material and bloom into sight.

The dirty-snowball hypothesis also accounted for erratic orbits like that of comet Encke. Elaborating on Friedrich Bessel's accurate—though neglected—surmise of more than a century earlier, Whipple conjectured that a comet's gas jets can hasten or retard its progress. In other words, the jets function much like the rocket thrusters that are used to keep interplanetary probes on course today: On a comet that travels in retrograde rotation—that is, one whose nucleus is rotating in a direction opposite that of its orbital path—the jets will fire forward, countering the comet's motion and thrusting it into a smaller orbit. The jets on a comet in direct (or prograde) rotation, by contrast, will fire rearward, reinforcing the comet's forward motion and boosting it into a larger orbit.

Stumbling points remained, however. Whipple had identified methane as a likely constituent of comet nuclei because the breakdown of methane by solar radiation would produce many of the chemical by-products that have been observed in scores of comets. Yet methane is also highly volatile: It therefore should not show up in the ejecta of any comet that had made more than a few trips past the Sun; still, scientists had repeatedly measured its presence (albeit indirectly) in some of the oldest comets known.

To explain this apparent departure of reality from theory, Belgian astronomers Armand Delsemme and Pol Swings proposed in 1952 that frozen water molecules might be acting to protect and preserve the comet's methane. Water, they knew, forms a neat three-dimensional lattice when it freezes, and it was possible that the myriad interstices so created could harbor molecules of methane. Nestled inside this protective grid, argued Delsemme and Swings, the methane would evaporate only when the water ice did.

Another mystery not fully explained by Whipple's theory was the mechanism that created the extended gas tails. The enigma was unraveled in 1951 by German astrophysicist Ludwig Biermann. As Eddington had established at the Oxford conference of 1908, a robust impetus was needed to create a comet's long, straight gas tail. Biermann hypothesized that this force was supplied by a "solar wind" of electrically charged protons and electrons streaming constantly from the Sun in all directions. As the solar-wind particles traveled outward at 900,000 miles per hour, said Biermann, they would blow the comet's coma into a gas tail that could grow to be 50 million miles long. (The gas tail is also called an ion tail, since the molecules of gas are ionized, or electrically charged.) The comet's other tail—its broad, curving appendage of dust—owed its shape to the mechanical pressure of sunlight, just as Arrhenius had suspected.

NEW CLOUD IN THE NEIGHBORHOOD
Over the next three decades, while astrophysicists refined their understanding of the solar wind, comet scientists worked to clarify their models of the dirty snowball and the Oort cloud. It was Biermann who initiated the veri-

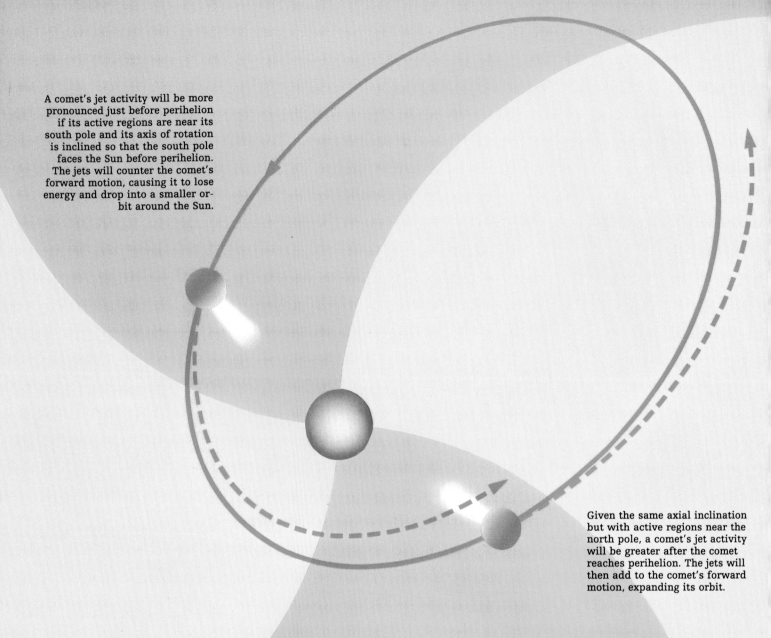

A comet's jet activity will be more pronounced just before perihelion if its active regions are near its south pole and its axis of rotation is inclined so that the south pole faces the Sun before perihelion. The jets will counter the comet's forward motion, causing it to lose energy and drop into a smaller orbit around the Sun.

Given the same axial inclination but with active regions near the north pole, a comet's jet activity will be greater after the comet reaches perihelion. The jets will then add to the comet's forward motion, expanding its orbit.

A FIERY ENCOUNTER WITH THE SUN

Whether a comet takes three years or 30 million years to draw its circle around the Sun, it spends the bulk of the journey as an inconspicuous chunk of ice and dust. But during its few months in the inner Solar System, the drab wanderer undergoes a spectacular transformation, emitting bursts of matter that not only modify its own trajectory but also give rise to the most striking portions of the comet's anatomy: its dual tails, one made up of dust, and the other of gas *(pages 34-35)*.

When an inbound comet reaches the orbit of Jupiter, about 485 million miles from the Sun, solar warmth begins to cause the water ice in the comet's nucleus to sublime, or transform directly from solid to gas. Along with particles of dust, the gas shoots out from the comet's sunward side in jets that act much like tiny rocket engines on a spacecraft, altering the comet's course *(above)*.

As the comet nears perihelion, its closest approach to the Sun, jet activity increases. The resulting spume of material forms an extensive coma—a glowing helmet that absorbs and reflects the Sun's rays and

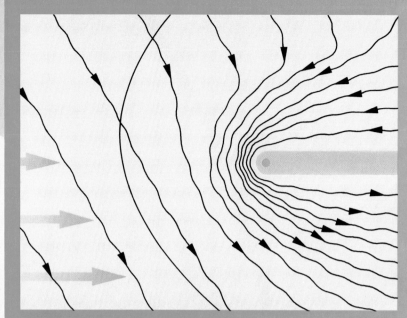

As the solar wind slows to subsonic speeds at the bow shock *(curved line)*, magnetic field lines pile up in front of the comet. Interactions among photons of solar radiation, particles of the solar wind, and the gas of the coma produce cometary ions, which are swept back into the ion tail *(above, right)*. Examples of these interactions are described below.

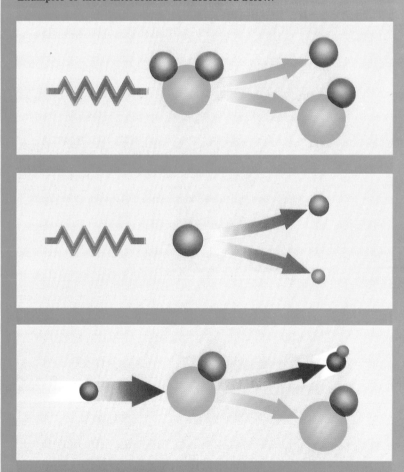

shields the cometary nucleus from view. Leading the nucleus by some 100,000 miles, the coma plows head-on into the solar wind, a plasma of charged solar gas that carries with it Sun-generated magnetic field lines, collectively known as the interplanetary magnetic field. Swirled into a spiral by the Sun's rotation, the interplanetary field is organized into alternating sectors of magnetic polarity (depicted here in light and dark yellow) directed toward or away from the Sun. Interactions between gas molecules in the coma and solar radiation and the solar wind *(right)* produce new molecules and yield the concentration of charged particles, or ions, that will form the comet's ion tail.

In one form of photodissociation *(above, top panel)*, a solar photon strikes a water molecule and splits it into a hydroxyl molecule (OH) and a hydrogen atom. Both contribute to a cloud that surrounds the visible coma. In photoionization *(middle panel)*, a photon can split a neutral hydrogen atom into a positively charged proton and a negatively charged electron, ions that help slow the solar wind and form the comet's tail. In one example of charge exchange *(bottom panel)*, a solar proton hits a neutral OH molecule, carries off an electron, and leaves behind a positively charged OH ion, which also adds to the ion tail.

As a comet nears perihelion, gas molecules in its coma fluoresce, emitting light visible from Earth. At about two astronomical units from the Sun, cometary ions captured by magnetic field lines carried in the solar wind flow around and behind the comet, forming the comet's ion tail. Since the solar wind emanates radially, the ion tail always streams directly away from the Sun, like a solar windsock; fluorescing carbon monoxide ions color it blue.

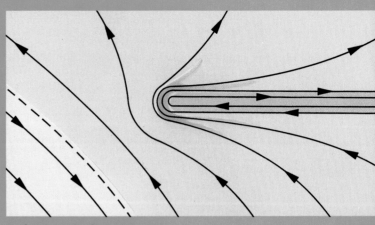

In the stages leading up to the process known as a disconnection event, a comet is approached by the neutral boundary *(dotted line)* between sectors of opposite polarity in the solar wind's magnetic field. Anchored to the Sun, magnetic field lines rotate with the Sun, completing one revolution every twenty-seven days. Since a comet at perihelion travels much more slowly, it is regularly overtaken by sector boundaries.

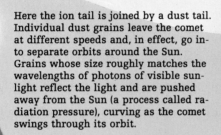

Here the ion tail is joined by a dust tail. Individual dust grains leave the comet at different speeds and, in effect, go into separate orbits around the Sun. Grains whose size roughly matches the wavelengths of photons of visible sunlight reflect the light and are pushed away from the Sun (a process called radiation pressure), curving as the comet swings through its orbit.

The comet's ion tail reaches its greatest length at perihelion, when it can stretch as far as 50 million miles into interplanetary space.

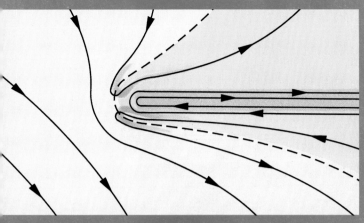

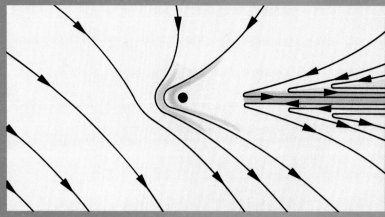

n the sector boundary catches up with the comet, magnetic field lines w and opposite polarity begin to pile up in front of the comet, thereby ing an area of instability. The lines on either side of the sector bound-reak and reconnect to one another until they are no longer wrapped nd the comet's head.

With the comet's coma surrounded by field lines in the new sector, the old tail pinches off and floats away into interplanetary space. The comet imme-diately starts to build a new ion tail, with a polarity corresponding to that of the new magnetic sector.

On the outbound leg of its journey, a comet flies tailfirst as radiation pres-sure and the solar wind push matter in the tails away from the Sun. Both tails begin to shrink as jet activity decreases with increasing distance from the Sun. When the comet encounters a sector boundary of the interplanetary magnetic field, a so-called disconnection event *(above diagrams)* can occur, in which the comet loses its ion tail.

Just past perihelion, as radiation pres-sure from the Sun pushes dust grains farther and farther out, the comet's dust tail can stretch out to several million miles in length.

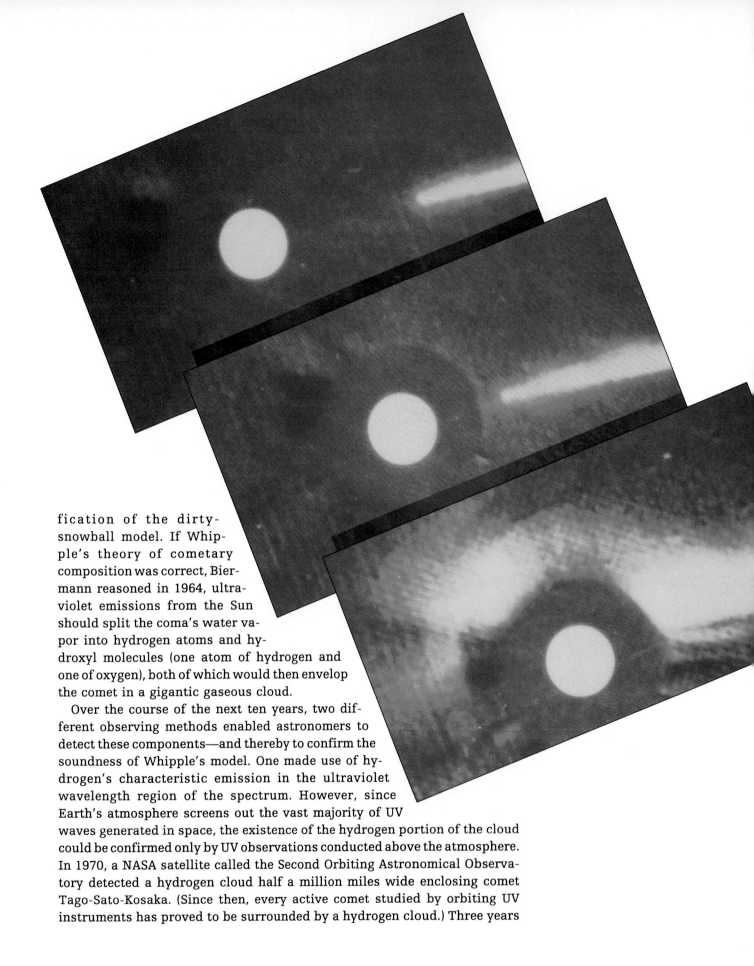

fication of the dirty-snowball model. If Whipple's theory of cometary composition was correct, Biermann reasoned in 1964, ultraviolet emissions from the Sun should split the coma's water vapor into hydrogen atoms and hydroxyl molecules (one atom of hydrogen and one of oxygen), both of which would then envelop the comet in a gigantic gaseous cloud.

Over the course of the next ten years, two different observing methods enabled astronomers to detect these components—and thereby to confirm the soundness of Whipple's model. One made use of hydrogen's characteristic emission in the ultraviolet wavelength region of the spectrum. However, since Earth's atmosphere screens out the vast majority of UV waves generated in space, the existence of the hydrogen portion of the cloud could be confirmed only by UV observations conducted above the atmosphere. In 1970, a NASA satellite called the Second Orbiting Astronomical Observatory detected a hydrogen cloud half a million miles wide enclosing comet Tago-Sato-Kosaka. (Since then, every active comet studied by orbiting UV instruments has proved to be surrounded by a hydrogen cloud.) Three years later, ground-based radio telescopes picked up radio wavelength emissions being given off by the hydroxyl cloud surrounding the nucleus of comet Kohoutek. The detection of hydrogen in the UV range and hydroxyl in the radio region proved that the water ice proposed by Whipple twenty years earlier is abundant indeed in comets.

Comet Kohoutek's appearance also gave astronomers the first of several clues that would lead them to confirm the prevailing theory of cometary origins: In 1973 and 1974, they identified within the comet a pair of carbon compounds—hydrogen cyanide and methyl cyanide—that were probably present in the presolar nebula from which all comets eventually took shape. Further examinations carried out by Delsemme in the 1980s established that the distribution of elements in the atmospheres of most comets roughly matches that of the Sun.

A COMET-FREE GAP

This finding reinforced the notion that comets evolved from the same primordial mix as the Solar System but pointed up a seeming discrepancy. Comets forming from the solar nebula should have come into existence everywhere that it was cold enough for their nuclei to congeal. Yet current theory set the inner boundary of the Oort cloud at 20,000 astronomical units from the Sun. To astronomer Jack Hills of New Mexico's Los Alamos National Laboratory, such an enormous comet-free gap was inexplicable. In 1981, Hills hit upon the idea of a second cometary nursery much closer. This inner cloud of comets, he suggested, extends from approximately the orbit of Neptune outward to the near edge of the Oort cloud and contains 200 times more comets than the Oort cloud itself. "The classical Oort cloud," Hills theorized, "may be just the halo of a much more extended comet distribution that reaches much closer in toward the Sun." Hills's inner cloud might even serve to restock the Oort cloud as the latter relinquishes its comets to the inexorable pull of passing stars.

An experiment conducted by a team of scientists in 1988 suggested that a third cometary reservoir may exist even closer to Earth than Hills's inner cloud. Physicist Martin Duncan of Queens University in Kingston, Ontario, working with astronomers Thomas Quinn and Scott Tremaine of the University of Toronto, constructed a computer model to test various origins for 121 short-period comets. To the trio's surprise, the scenario that depicted short-period comets as arising in the Oort cloud proved a poor match with observation; comets ejected from the Oort cloud in the computer model did not wind up in the low-inclination orbits that short-period comets are known to follow. The researchers then fashioned a second computer model, this one incorporating Gerard Kuiper's original notion of a flattened inner belt of comets girding the Solar System just beyond the orbit of Neptune. The revised program then yielded a distribution of short-period comets within the Solar System that was strikingly similar to the one that had been observed: Jettisoned from the Kuiper belt by the gravitational perturbations of Neptune—

A series of images produced by a coronagraph on the solar research satellite SOLWIND on August 30 and 31, 1979, shows comet Howard-Koomen-Michels plunging to a fiery death in the Sun's atmosphere. The coronagraph's occulting disk, which blocked the Sun (drawn in for clarity) and a portion of the inner corona, hides the comet's final moments. In the last view, only remnants of the dust tail emerge on the Sun's opposite side. Howard-Koomen-Michels was probably one of the so-called sun-grazing comets, several of which follow similar orbits and may be pieces of a larger comet that fragmented during a close solar passage.

THE BREAKUP
OF COMET WEST

A vision of celestial splendor, comet West graced predawn skies in the spring of 1976. The comet's two tails can be distinguished by their different colors in the image at near right: Sunlight-reflecting particles in the shorter dust tail glow orange and pale yellow, while energetic gas molecules in the more extensive plasma tail radiate at the blue end of the visible spectrum. After passing within 19 million miles of the Sun, the comet's nucleus broke into four pieces, as shown in the sequence of photographs at far right, taken over the course of ten days in March. Each fragment later grew its own tail.

or even of particularly large fellow comets—the simulated comets wound up in low-inclination, short-period, prograde orbits that matched the paths of the observed comets.

A COMETARY HOMECOMING

The slated reappearance of Halley's comet in 1986 gave scientists the world over an ideal opportunity to examine the most famous comet of all—and the brightest of the short-period breed. Although comets had passed close to Earth before—comet Lexell sped by at a distance of just 1.4 million miles in 1770, for example—rarely had scientists been afforded the chance to peer at them from close range, above Earth's atmosphere.

All around the globe, Halley's renown helped astronomers secure governmental support for deep-space probes designed to rendezvous with the comet near perihelion and scrutinize it in optical and nonoptical wavelengths. The notable exception to this scientific mobilization was the United States, where NASA—under budgetary pressure from the Reagan administration—could not muster the funds to send a probe to Halley. American engineers and astronomers therefore scrambled to berth their instruments on the expeditions of other nations.

Halley's comet and the international armada of space probes scheduled to greet it took center stage in the spring of 1986. The Soviet ground controllers in charge of the two Vega spacecraft were assisted by a coalition of scientists from nine nations. The European Space Agency's entry, Giotto (named after the Italian painter who had captured the first accurate images of Halley some 700 years earlier), also represented a team effort, and it was ESA's first interplanetary probe. Japan's Suisei and Sakigake emissaries were likewise that country's inaugural deep-space craft.

This combined fleet encountered a comet radically different from the one scientists had expected *(pages 42-49)*. Halley's nucleus was larger, more irregular, less dense, less homogeneous, less volatile, finer grained, and darker than anticipated, and it rotated in a wobbly fashion every 2.2 days. As these facts sank in, astronomers hastily amended their views of cometary nuclei. The black coloration of Halley's nucleus, they deduced, was imparted by a fluffy blanket of some carbon-based, tarlike substance that covered about 90 percent of the surface.

Some scientists believe that the molecules of this substance are configured to form a highly porous material, which would account for the fact that Hal-

39

ley's comet absorbs 97 percent of the light it receives; light rays striking the nucleus would become literally trapped in the mesh of pores. Others have proposed that the comet is made up of hundreds of porous boulders, all cemented together with a mixture of ice and snow known facetiously as Whipple glue. What is certain is that pockets of volatile ices lie here and there just beneath the surface of the nucleus. This is essentially Whipple's dirty snowball, ready to vaporize as it nears the Sun and to send its gas jets streaming toward that star.

In these jets, Giotto's instruments determined, are the apparent fragments of large carbon-based molecules. Using a device called a positive ion cluster composition analyzer on board the craft, Giotto's controllers sampled tiny amounts of the tens of tons of matter ejected by Halley's nucleus each second. Among their finds was a complex molecule called polyoxymethylene—a molecular shard of a larger formaldehyde polymer, or chain, that forms a light-weight mesh upon coming in contact with minuscule grains of interstellar dust. The discovery buttresses the notion that comets condensed from clouds

AN INTERNATIONAL FLEET OF PROBES

When Halley's comet returned from the depths of the Solar System in 1986, it was greeted by a handful of spacecraft launched from around the world to investigate the famed visitor. Two Japanese satellites took measurements from a distance: Suisei gathered information from 100,000 miles ahead of the nucleus, well inside the shock front, the boundary where cometary ions meet the streaming particles of the solar wind; more than four million miles farther ahead, Sakigake concentrated on the makeup of the wind itself. Three other craft—the Soviet Union's Vega 1 and Vega 2 and the European Space Agency's Giotto—braved the storm of dust particles in Halley's coma for closeup studies. Despite protective measures, all three probes sustained heavy damage yet returned significant new data on the composition of the nucleus and the dynamics of the dust and gas surrounding it. Giotto, the best performer, ventured within 400 miles of the nucleus to return the first clear images of Halley's solid core.

Europe: Giotto
Protected from flying dust particles by an aluminum shield *(top),* Giotto housed eleven different experiments

of gas and dust as the Solar System took shape some 4.6 billion years ago.

Other riddles of cometary composition and behavior may be cleared up during NASA's Comet-Rendezvous, Asteroid-Flyby (CRAF) mission, scheduled for launch in 1995. Somewhere between Mars and Jupiter, near the end of its five-year journey, the probe will meet up with comet Kopff, which rounds the Sun every six years. CRAF's array of sophisticated instruments—cameras, spectrometers, and a miniature electron microscope capable of scrutinizing individual particles of dust—will enable scientists to submit Kopff to the most thorough cometary examination ever.

The highlight of the mission will come when CRAF hurls a five-foot-long, golf-tee-shaped titanium harpoon into Kopff's nucleus. The rocket-guided "penetrator-lander" will lodge in the icy surface, whereupon a spectrometer, a calorimeter, and gas analyzers inside the harpoon will read the temperature of the comet's nucleus and precisely determine its composition. Fortified with this information, astronomers should be able to draw a more accurate picture of the origin—and ultimate fate—of comets within the Solar System.

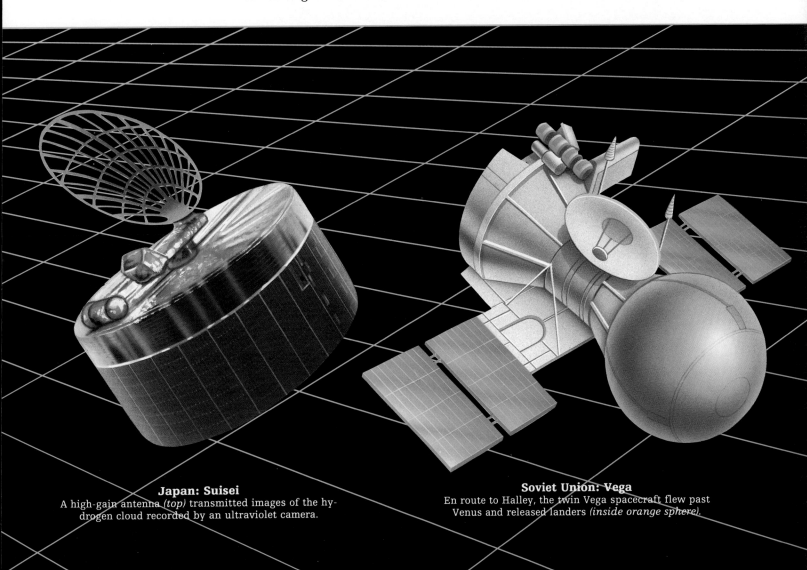

Japan: Suisei
A high-gain antenna *(top)* transmitted images of the hydrogen cloud recorded by an ultraviolet camera.

Soviet Union: Vega
En route to Halley, the twin Vega spacecraft flew past Venus and released landers *(inside orange sphere)*.

MISSION TO HALLEY'S COMET

English astronomer and mathematician Edmond Halley's deduction, in 1695, that three comets recorded in 1682, 1607, and 1531 were in fact one repeat visitor marked the beginning of cometary science—the first recognition that comets are members of the Solar System, orbiting the Sun just as planets do. When he subsequently predicted that the comet that now bears his name would return in late 1758 or early 1759, he planted the seeds of the popular and scientific anticipation that has attended Halley's appearances ever since. During its first twentieth-century visit, in 1910, the comet dazzled earthbound comet watchers when its brilliant tail, millions of miles long, slashed its way across the night sky.

Seventy-six years later, the technology of space travel gave scientists a chance to view Halley more intimately. A flotilla of five robot spacecraft, each bristling with sensors, headed out to meet the comet as it crossed Earth's orbit in March 1986. The missions were a testament to international cooperation. As described on the following pages, one craft, the European Space Agency's Giotto (left), used positioning data from two of the others to zero in on Halley's core, or nucleus, and information from all five probes was widely shared.

Because comets are geologically inactive and spend most of their time far from the corrosive heat of the Sun, scientists have long regarded them as orbiting time capsules, potential containers of primordial stuff left over from the formation of the Sun and its planets some 4.6 billion years ago. The expedition to comet Halley at least confirmed the existence of a solid nucleus and offered some intriguing hints about conditions in the early life of the Solar System.

Since Halley's clockwise orbit (orange) is inclined 162 degrees to the ecliptic—the plane of Earth's orbit (blue)—its trajectory brings it up and over the ecliptic plane as it swings around the Sun. Giotto (red) and the other spacecraft, orbiting counterclockwise, like Earth, met the comet in March 1986, just as it crossed Earth's orbit a second time, flying tailfirst out of the inner Solar System.

THE ART OF INTERCEPTION

Calculating the five space probes' close encounters with Halley's comet in 1986 was a tricky proposition, for a comet's orbital path grows increasingly unpredictable as it nears the Sun. Irregular jets of dust and gas burst from its sunward surface, changing its course just as steering rockets alter the direction and speed of a spacecraft. Because these forces are nongravitational, a comet's motion cannot be precisely determined in advance (although computer models can give some indication of the jets' effects). Instead, ground controllers for the Halley flyby had to guide their craft to an absolute point in space based on more than 4,000 positional measurements made from Earth between October 1982, when the comet was sighted just beyond Saturn's orbit, and March 1986, when it crossed Earth's. To bring Giotto as close as possible to Halley's nucleus

on the sunward side required a globe-spanning cooperative effort, dubbed Pathfinder. The European Space Agency used NASA's Deep Space Network to relay data on the comet's actual position from the two Soviet craft, Vega 1 and Vega 2, and then adjusted Giotto's approach accordingly (below).

Because Halley travels around the Sun clockwise instead of counterclockwise as the planets do, the comet and each of the probes in turn would shoot past each other like cosmic bullets. To conserve fuel and to ensure that the probes would catch Halley during its most active phase, the craft were launched so as to meet Halley at the second of two "nodes," where the comet's orbit intersects the ecliptic plane (opposite). By this time, Halley would have rounded the Sun, heading back out for its long sojourn in the depths of space.

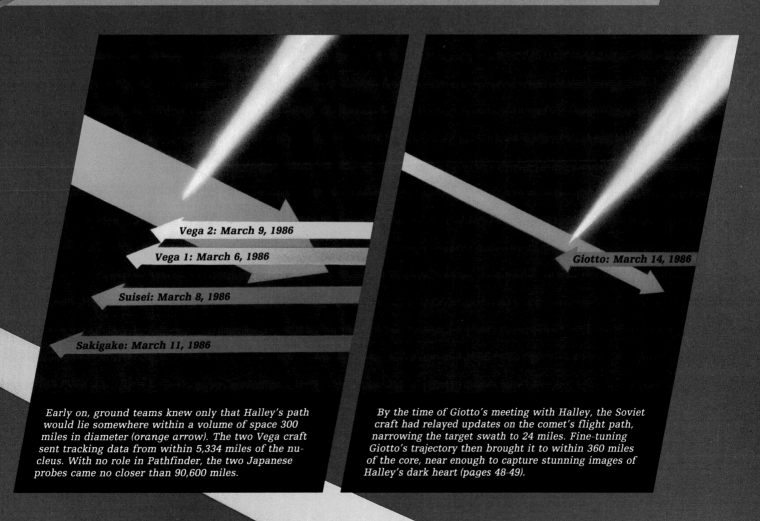

Vega 2: March 9, 1986

Vega 1: March 6, 1986

Suisei: March 8, 1986

Sakigake: March 11, 1986

Giotto: March 14, 1986

Early on, ground teams knew only that Halley's path would lie somewhere within a volume of space 300 miles in diameter (orange arrow). The two Vega craft sent tracking data from within 5,334 miles of the nucleus. With no role in Pathfinder, the two Japanese probes came no closer than 90,600 miles.

By the time of Giotto's meeting with Halley, the Soviet craft had relayed updates on the comet's flight path, narrowing the target swath to 24 miles. Fine-tuning Giotto's trajectory then brought it to within 360 miles of the core, near enough to capture stunning images of Halley's dark heart (pages 48-49).

INTO THE DUSTY ENVELOPE

The robot investigators sent to meet comet Halley carried a battery of instruments, including cameras, magnetometers, particle detectors, and several types of spectrometers. Most of the craft were designed to penetrate the comet's coma (opposite)—the roughly spherical, expanding cloud of gas and dust that defines the comet as seen from Earth. By working backward from an analysis of gas and dust in the coma, scientists hoped to deduce the composition of the nucleus. They also wanted to study the interaction of the coma with the solar wind and the interplanetary magnetic field.

Giotto found that the comet's coma is about 80 percent water molecules, with most of the balance consisting of molecules of carbon monoxide. This finding confirmed at least one aspect of the early "dirty-snowball" model of comets, which theorized that the nucleus is largely water ice.

The makeup of Halley's dust was something of a surprise. From ground-based studies of many comet atmospheres, astronomers had made the determination that cometary gases contained less than one-third the carbon that is found in the rest of the Solar System. This led scientists to believe that the comet's dust would also be low in carbon. However, Giotto found a type of dust particle, which scientists labeled CHON, that may contain proportionately as much carbon as the Sun. The discovery not only accounts for the missing atmospheric carbon but also bolsters the notion that comets are remnants of the stuff of stars.

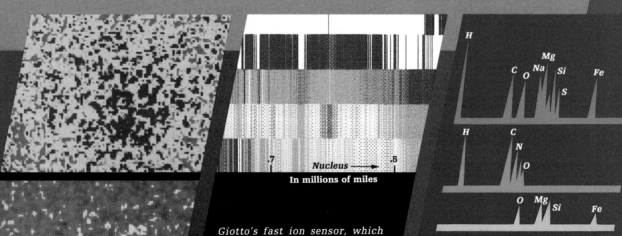

Nucleus ⟶

In millions of miles

Suisei recorded a 2.2-Earth-day cycle of fluctuations in ultraviolet emissions by neutral hydrogen in Halley's corona. As active areas on the nucleus rotated into and out of sunlight, high jet action led to strong emissions (red, top image), diminished action to weak emissions (green, bottom image)—a

Giotto's fast ion sensor, which measured the energy and concentration of charged particles, or ions, in Halley's coma, helped define the bow shock region, the boundary between the corona and an area called the magnetosheath. As indicated above, ion counts began rising (yellow) at about 800,000 miles from the nucleus, marking the onset of the bow shock. At the time of Giotto's passage, the bow shock was 24,000 miles thick. As magnetic field lines in the solar wind began picking up increasing amounts of cometary ions, the speed of the solar wind dropped from 180 miles per second to 156, and the concentration of ions increased from one per cubic centi-

Giotto's dust mass spectrometer revealed three types of dust particles. One type (top graph) was found to contain a mix of elements matching no particular mineral, but another type (bottom graph) contained elements found in silicate rocks. Most intriguing was the dust depicted in the middle graph: A mixture of carbon, hydrogen, oxygen, and nitrogen (CHON), the dust contains carbon in proportions that match the known carbon abundance of stars,

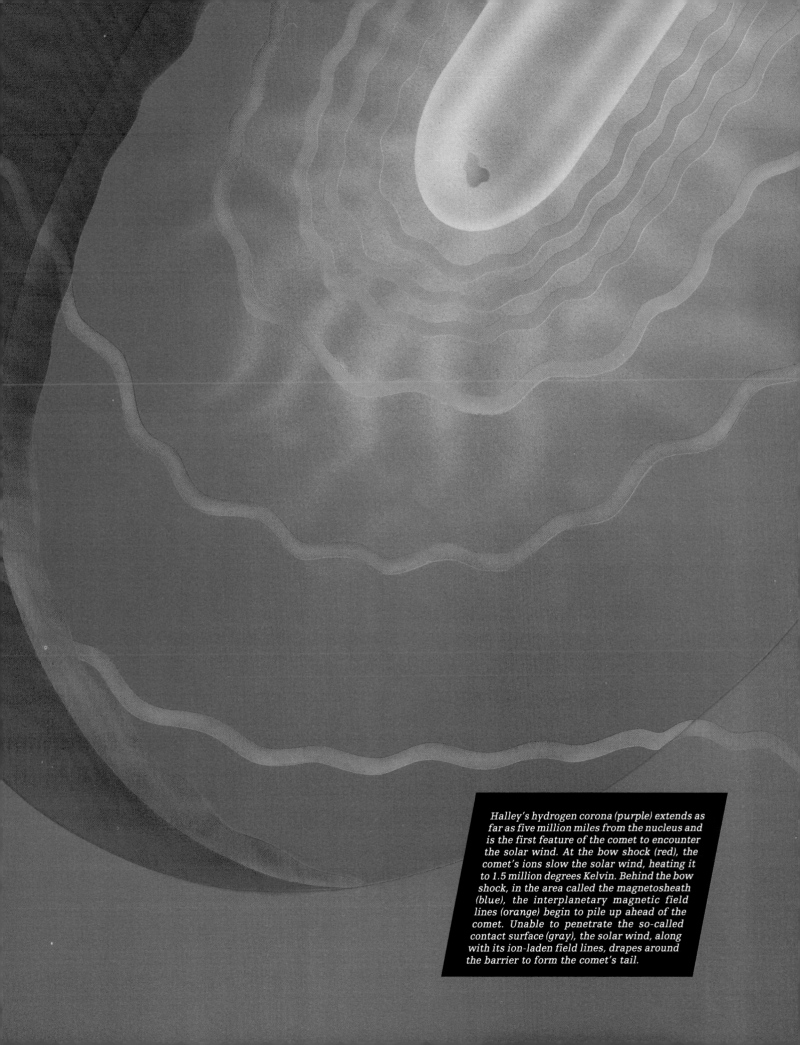

Halley's hydrogen corona (purple) extends as far as five million miles from the nucleus and is the first feature of the comet to encounter the solar wind. At the bow shock (red), the comet's ions slow the solar wind, heating it to 1.5 million degrees Kelvin. Behind the bow shock, in the area called the magnetosheath (blue), the interplanetary magnetic field lines (orange) begin to pile up ahead of the comet. Unable to penetrate the so-called contact surface (gray), the solar wind, along with its ion-laden field lines, drapes around the barrier to form the comet's tail.

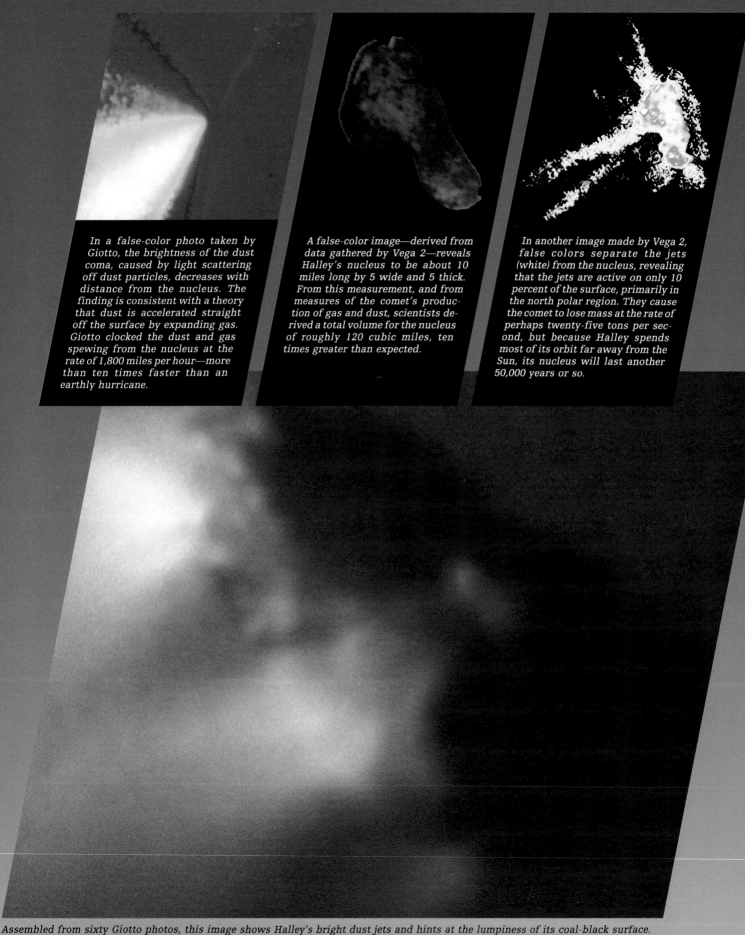

In a false-color photo taken by Giotto, the brightness of the dust coma, caused by light scattering off dust particles, decreases with distance from the nucleus. The finding is consistent with a theory that dust is accelerated straight off the surface by expanding gas. Giotto clocked the dust and gas spewing from the nucleus at the rate of 1,800 miles per hour—more than ten times faster than an earthly hurricane.

A false-color image—derived from data gathered by Vega 2—reveals Halley's nucleus to be about 10 miles long by 5 wide and 5 thick. From this measurement, and from measures of the comet's production of gas and dust, scientists derived a total volume for the nucleus of roughly 120 cubic miles, ten times greater than expected.

In another image made by Vega 2, false colors separate the jets (white) from the nucleus, revealing that the jets are active on only 10 percent of the surface, primarily in the north polar region. They cause the comet to lose mass at the rate of perhaps twenty-five tons per second, but because Halley spends most of its orbit far away from the Sun, its nucleus will last another 50,000 years or so.

Assembled from sixty Giotto photos, this image shows Halley's bright dust jets and hints at the lumpiness of its coal-black surface.

A CLOSEUP OF THE NUCLEUS

Enveloped by its dusty coma, the tiny nucleus of Halley's comet is invisible from Earth, even with the most powerful optical telescopes. On nearly all counts—size, density, and color—Halley's hidden core was a cornucopia of surprises. First, it appears to be four times longer than previously estimated, roughly the size of the island of Manhattan. It also has an irregular terrain, smooth in some sectors, hilly in others, signs that the nucleus formed from several smaller bodies.

The composition of Halley's nucleus also turned out to be far more complicated than the dirty-snowball model suggested. The probes revealed a surface as black as soot, reflecting only three percent of the light that hits it. Halley's comet is thus not only one of the darkest objects residing in the Solar System but also considerably hotter than scientists had expected it to be. Spectrographic measurements indicated a surface temperature of as much as 400 degrees Kelvin, 80 to 100 degrees too warm for the surface to be evaporating bare ice. Comet specialists thus drew the conclusion that Halley has a crust of some nonevaporating material. The unevenness of such a covering would also help explain why the comet's jets occur so irregularly over the surface.

Given a mixture of ice and dust, scientists had expected the cometary material to have a density a little higher than that of water ice. Instead, with the newfound size of the nucleus, they determined it to be extremely porous: With an average density one-tenth to one-fourth that of ice, the comet may be more like a dirty snowdrift than snowball.

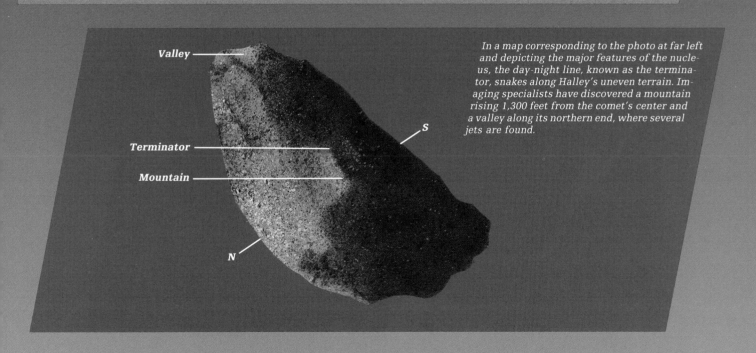

Valley

Terminator

Mountain

S

N

In a map corresponding to the photo at far left and depicting the major features of the nucleus, the day-night line, known as the terminator, snakes along Halley's uneven terrain. Imaging specialists have discovered a mountain rising 1,300 feet from the comet's center and a valley along its northern end, where several jets are found.

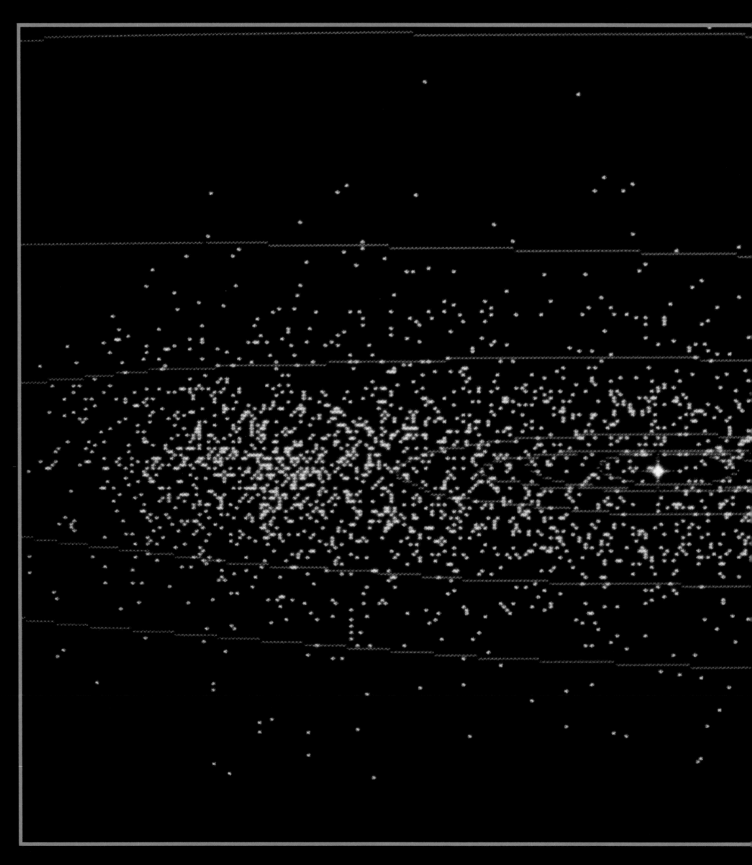

PLANETS

Like a swarm of fireflies, thousands of miniature planets called asteroids circle the Sun in this computer simulation of the Solar System. Most inhabit the 350-million-mile gap between Mars (fourth orbit from center) and Jupiter (fifth orbit), whose powerful gravity kept the rocky bodies from accreting into one large planet as the Solar System formed.

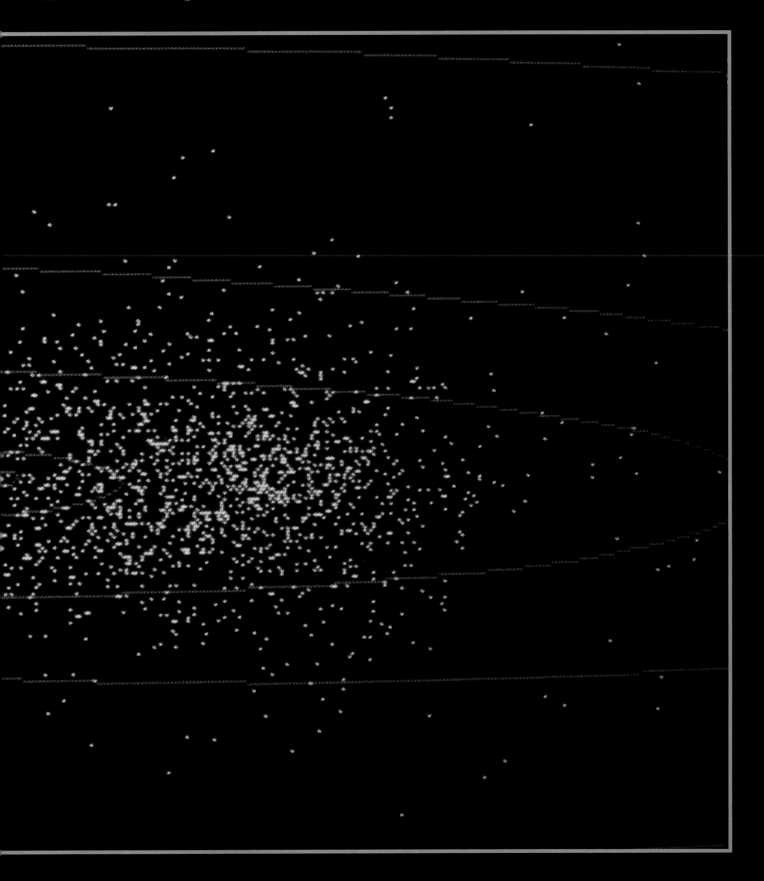

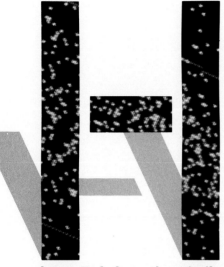urtling through space at 46,000 miles an hour, an asteroid crossed Earth's orbit on March 23, 1989, just 400,000 miles ahead of the planet—only 160,000 miles farther than the Moon. The intruder then sped harmlessly on its way; as a matter of fact, no one realized it had gone by until several days later. Had the rocky missile followed a slightly different path, however, human history would have been altered dramatically. The object was about a quarter of a mile in diameter and weighed some 50 million tons. Its collision with Earth would have been catastrophic, if not apocalyptic. At sea it might have raised tsunamis hundreds of feet high that would have obliterated the coastal areas they washed over. On land the object easily could have destroyed a city the size of New York and thrown up clouds of dust and debris that would have blocked the Sun's light and devastated life on Earth.

"On the cosmic scale of things," said geologist and part-time astronomer Henry Holt of the U.S. Geological Survey in Flagstaff, Arizona, "that was a close call." Holt spotted the asteroid about a week after this near miss. As part of a NASA-funded asteroid survey, he was poring over photographs taken on March 31 with the eighteen-inch Schmidt telescope at California's Palomar Observatory when he detected a telltale streak of light on a pair of images viewed through a stereo microscope. For the next several nights, he and a number of fellow astronomers used the Schmidt to track the rapidly receding object and established that it travels from a point inside the orbit of Mars on the outer extreme of its journey to somewhere inside Earth's orbit as it approaches the Sun. The group's calculations revealed that Earth and the asteroid had passed nearly the same point in space just six hours apart.

The Minor Planet Center, located since 1978 at the Harvard-Smithsonian Center for Astrophysics in Cambridge, Massachusetts, gave the asteroid a provisional name, 1989 FC: The F indicates that the discovery occurred in the sixth half-month of 1989, and the C denotes that this was the third new asteroid to be found during that period. If 1989 FC is seen again on one more pass, tradition declares that Holt may give it a permanent name. Whatever designation he chooses may ultimately acquire a certain notoriety: If Holt's orbital calculations are correct, 1989 FC returns to the inner Solar System every year, and it stands a good chance of someday hitting Mars, Venus, the Moon, or Earth.

With more than one million asteroids orbiting the Sun between Mars and Jupiter, objects like 1989 FC strike their bigger siblings in the Solar System quite often. The record of violence is written in the pocked faces of Mars, the Moon, and Mercury, which bear the scars of an era billions of years ago when asteroids swarmed the interplanetary expanses in far greater abundance than they do today. Even Earth retains indelible signs of a calamitous past: Geologists have identified more than 120 impact craters around the world, and they have also uncovered chemical and mineralogical evidence suggesting that an asteroid wiped out the dinosaurs and many other species when it slammed into the planet some 65 million years ago *(pages 121-133)*.

Despite their obvious significance to terrestrial life, asteroids were once relegated to the minor leagues of astronomy. As advances in spectroscopy and astrophysics in the first half of the twentieth century revealed a wide range of cosmic exotica, from the birth and death of stars to the violent intermingling of passing galaxies, asteroids simply seemed to get in the way. One astronomer remembered the 1950s as a time when most major observatories regarded asteroids as "the vermin of the sky." A few decades later, however, NASA began planning missions beyond the Moon, and asteroids came back into scientific vogue. Space engineers contemplated them as viable targets for future landings; geologists considered dispatching human or robotic prospectors to search them for minerals. Even cosmogonists caught the asteroid bug, seizing on the minor planets as rich—and relatively unchanged—sources of information about the origins of the Solar System.

PURSUING A PHANTOM

Although asteroids are clearly long-time residents of the Solar System, they are relatively new to astronomy. The sky watchers who spotted the first asteroids in the early nineteenth century were looking for something else entirely: a mysterious missing planet that many believed awaited discovery somewhere beyond Mars.

The idea had originally been suggested by German astronomer Johannes Kepler, who subscribed to the view that an ineffable unity of design governed the Solar System. In 1596, while contemplating the planets' relative distances from the Sun, Kepler began to suspect that the 350-million-mile stretch of space separating Jupiter from Mars was simply too vast to be unoccupied. "Inter Jovem et Martem interposui planetam," he announced: "Between Jupiter and Mars I place a planet."

By the mid-1700s, there was still no sign of Kepler's postulated planet, but a small band of European astronomers were working hard to find it. In 1766, for example, Kepler's countryman Johann Titius devised a formula that confirmed the mathematical likelihood of a missing planet. Titius had observed that the four inner planets orbit the Sun at distances—expressed in astronomical units (AU)—arrived at by a formula that included an exponentially increasing variable, N. A planet's distance from the Sun, he calculated, equals $.4$ AU $+ (.3$ AU $\times N)$, where N doubles for each planet out: 0 for Mercury,

1 for Venus, 2 for Earth, 4 for Mars, and so on. A snag arose, however, when Titius inserted the value 8 for the next known planet, Jupiter. Instead of producing Jupiter's observed distance—5.2 AU from the Sun—the formula indicated that an undiscovered planet should orbit at 2.8 AU. Except for that one deviation, the equation remained valid: An N value of 16 yielded the actual distance of Jupiter, while an N value of 32 resulted in the correct position (10 AU) for Saturn.

As it happened, another German astronomer, Johann Bode, hit upon the identical formula at the same time. So effectively did Bode promote the equation that it became known as the Titius-Bode law—and ultimately, simply Bode's law. The theory was upheld—or so it seemed—in 1781, when William Herschel discovered the Solar System's seventh planet, Uranus, orbiting precisely where Bode's law said it should (that is, at 19.6 AU) when N equals 64. (The proof crumbled, however, when Neptune and Pluto were found in 1846 and 1930, respectively; the eighth and ninth planets orbit much closer to the Sun than Bode's law permits.)

The event launched a young Hungarian noble on a quest for the world that should lie in the large gap between Mars and Jupiter. In 1785, Baron Franz von Zach reported that he had used Bode's math to calculate a more precise orbital distance of 2.82 AU for the missing planet; von Zach had also computed a host of additional details—including the planet's presumed orbital period, its inclination (the angle between its orbital plane and the ecliptic plane), and

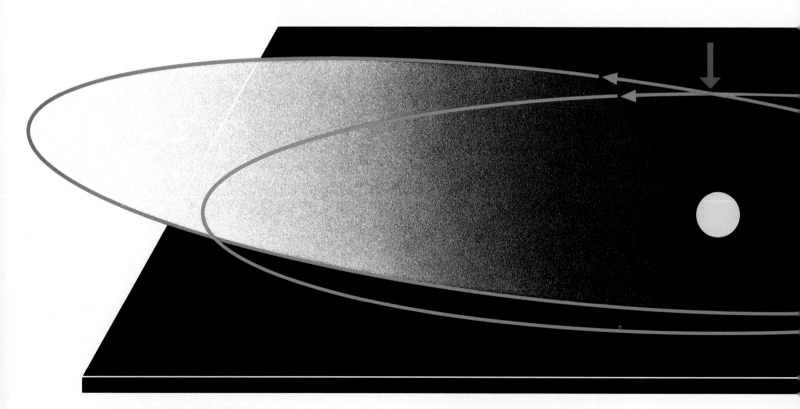

its orbital eccentricity, or degree of deviation from a perfect circle—that would be required to locate it. All that was needed now, said von Zach, was for someone to start looking.

Fifteen years passed without a glimpse of the mystery planet, so von Zach decided to organize an international search party. On September 11, 1800, he divided the night sky into twenty-four segments and assigned each to a colleague or friend, whose duty was to observe that region at every opportunity until the planet was found. (With the exception of two observers, von Zach noted later, everyone accepted the task "with delight.") German astronomer Johann Schroeter agreed to chair the group, which came to be known as the "celestial police." Shortly after the posse assembled at Schroeter's observatory in Lilienthal near Bremen, however, its quarry was inadvertently apprehended by a lone observer elsewhere.

Giuseppe Piazzi, a Sicilian monk who served as director of the Palermo Observatory, was correcting an inaccurate star catalog in late December of 1800 when he noticed something odd. On New Year's Eve, Piazzi detected a light shining in the constellation Taurus that did not appear on his star chart. Not only was the light unrecorded, but it was moving; Piazzi therefore took it to be a new comet. As he tracked the object over the next three weeks, however, he observed that its image was sharper and its progress was slower than those of a typical comet. On January 24, 1801, Piazzi aired his doubts in a letter to Barnaba Oriani, a fellow astronomer living in Milan: "The fact

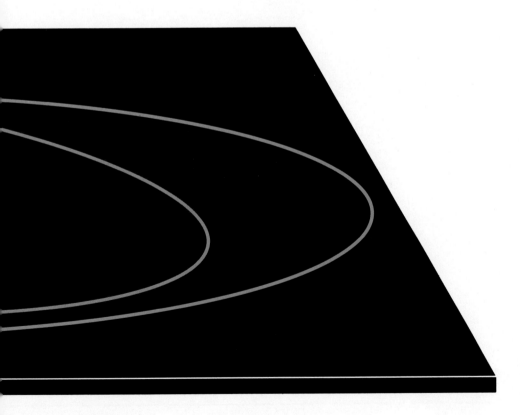

In March of 1989, Earth narrowly escaped a cataclysmic encounter with a quarter-mile-wide asteroid dubbed 1989 FC. As shown at left, the asteroid's orbit *(orange)* inclines at a narrow angle to the ecliptic plane *(black)*, nearly crossing Earth's path twice. On March 23, the two bodies passed within 400,000 miles of each other *(arrow)*.

that the star is not accompanied by any nebulosity, and that its movement is very slow and rather uniform, has caused me many times to seriously consider that perhaps it might be something better than a comet."

Piazzi also conveyed his misgivings to Bode, who credited the Italian with having discovered the missing planet. Piazzi named the body Ceres Ferdinandea to honor both Sicily's patron goddess, Ceres, and its king, Ferdinand. Other astronomers, judging it crass to title a planet after a living ruler, shortened the designation to Ceres.

Verifying Piazzi's discovery posed a problem. For one thing, Piazzi resisted sharing his positional data with other astronomers, because he wanted the honor of publishing the first orbital details himself. Before he could do so,

THE ASTEROID HUNTERS

The scientific detectives who first tracked the asteroids were acting on a false lead—Johannes Kepler's claim in 1596 that a major planet must exist in the vast space between Mars and Jupiter. Interest reached a peak two centuries later when Kepler's speculation was seemingly confirmed by a mathematical formula. The investigation soon took a surprising turn, however, as observers located not one large body but many smaller ones. A prominent theory maintained that these were the remnants of an exploded planet. But as astronomers refined their techniques, they realized that asteroids, far from being the humble offspring of a single parent, constitute an ancient and sprawling clan, some of whose members pose a threat to Earth itself.

1596 Assuming that the Solar System must be orderly in design, the German astronomer Johannes Kepler posited the existence of a planet in the gap between Mars and Jupiter.

1772 German Johann Bode spurred interest in the missing planet by predicting its distance from the Sun according to a formula first publicized by fellow German Johann Titius.

1800 Baron Franz von Zach of Hungary organized a band of astronomers known as the celestial police to find the planet.

1801 Acting independently, Sicilian monk Giuseppe Piazzi was the first to observe an orbiting body between Mars and Jupiter. He named it Ceres.

however, the heavens conspired against his cause: In the spring of 1801, as Ceres's orbit took it nearer and nearer the Sun relative to the line of sight from Earth, the object became unobservable for several months, making it difficult, if not impossible, for anyone to calculate the orbit.

Then a German mathematical genius named Karl Friedrich Gauss intervened. Earlier that year, at the age of twenty-four, Gauss had published *Disquisitiones Arithmeticae*, one of the greatest mathematical treatises in history. By September he had learned of Piazzi's discovery, and he took up the puzzle of Ceres's orbit as a diverting challenge. Using the scant information provided by Piazzi, Gauss reconstructed Ceres's complete orbit with a complex analytical technique of his own devising.

In November, Gauss announced that Ceres circles the Sun at a distance of 2.77 AU, slightly less than the 2.82 AU that von Zach had predicted. He also determined the object's orbital eccentricity as .08, meaning that its orbit was just slightly elliptical, and its inclination as a moderate 11 degrees. (Von Zach's prediction for the eccentricity of Ceres's orbit was close at .14, but his estimate of its inclination—only 1.5 degrees—fell wide of the mark.) When astronomers used Gauss's calculations to aim their telescopes, they succeeded in relocating Ceres; von Zach recovered it on December 7, and German astronomer Heinrich Wilhelm Olbers—another member of the celestial police—spotted it on December 31.

Not long after Ceres was retrieved from its orbital oblivion, sky gazers

1802 After Pallas was sighted, British astronomer William Herschel dubbed it and Ceres "asteroids"—for "starlike."

1802 Gauss's countryman Heinrich Wilhelm Olbers discovered the first of Ceres's neighbors, Pallas. He speculated that both were vestiges of a planet that exploded.

1801 German mathematician Karl Gauss drew on Piazzi's data to calculate the orbit of Ceres, setting the stage for systematic observations that revealed other objects in Ceres's vicinity.

discovered that the "missing planet" had some traveling companions. In March of 1802, as Olbers was observing Ceres, he identified a second minor planet in the same area of the sky and named it Pallas. (At about this time, astronomers agreed that these small, bright objects were not planets. At the suggestion of English astronomer William Herschel, they were named asteroids—meaning "starlike." Later observers have advocated the use of "planetoids" or "minor planets" as more accurate labels.)

Olbers theorized that both Ceres and Pallas had come from a single source—namely, a large planet that had exploded, leaving fragments that had continued orbiting the Sun in the path of their erstwhile parent. Corroboration for this theory seemed to come two years later, when another member of the celestial police, Karl Harding, found a third asteroid, which was dubbed Juno. Then, in March of 1807, Olbers discovered a fourth, Vesta, which to this day is the brightest of all.

Olbers's theory for the origin of asteroids would come to be discounted by virtually all astronomers, but in fact there is a mystery planet lurking in the lives of the asteroids. Its name is Jupiter. That the largest of all planets should have an effect on the smallest came as no surprise. Newton had established

1932 Hot on the heels of Delporte's find, German Karl Reinmuth detected Apollo, the first asteroid that is known to cross Earth's orbit.

1866 American mathematician Daniel Kirkwood linked unoccupied orbits in the main asteroid belt to Jupiter's gravity and orbital frequency.

1891 German astronomer Maximilian Wolf inaugurated time-exposure photography to track asteroids—and discovered more than 200 in his life.

1918 Japanese astronomer Kiyotsugu Hirayama began bringing order to the seemingly chaotic profusion of asteroids by classifying them into families.

1932 Belgian Eugène Delporte kindled interest in Earth-approaching asteroids when he discovered Amor, which came within 10 million miles.

the effects of gravitational forces on orbital motion more than a century earlier. However, the extent of Jupiter's influence on asteroids remained obscure until astronomers had collected more data. Such data were slow in coming: After Olbers found Vesta, no new asteroids were discovered for thirty-eight years.

HANDICAPS

The reason for the observational drought was twofold. First, with the search for the missing planet now at an end, few observers expected to find any additional asteroids. Second, star catalogs and maps of the era lacked the detail necessary for astronomers to distinguish minor planets fainter than Juno. Like Giuseppe Piazzi, any observer who wished to discover an asteroid in the early nineteenth century had to compare the position and magnitude of stars glimpsed in a telescope with those shown on a hand-drawn chart. A "star" that appeared in the sky but not on the chart then had to be monitored for several hours to determine whether it was moving with respect to the fixed stars; if so, the new object was very likely an asteroid. Reliable and complete star charts were essential to this process.

1973 Eugene Shoemaker began tracking Earth approachers at Palomar Observatory in California. His wife, Carolyn, an expert spotter in her own right, discovered half a dozen such intruders by 1985.

1983 Asteroid tracking was automated when Tom Gehrels attached a device to the 36-inch telescope at Kitt Peak Observatory that digitized images for computer analysis.

1976 American Eleanor Helin sighted the first Earth crosser with an orbit closer to the Sun than Earth's on average. She dubbed the new type Aten after an ancient Egyptian Sun god.

In 1845, a German postmaster and amateur astronomer named Karl Hencke used the recently published—and much more detailed—star charts of the Berlin Academy of Science to spot a fifth asteroid, which he named Astraea. Hencke's discovery inspired amateurs and professionals alike to resume the quest for minor planets, so that by 1850 the number of known asteroids had doubled to ten.

The most successful of these searchers was Hermann Goldschmidt, a German artist living in Paris. While painting a portrait of Galileo in 1852, he happened to attend an astronomy lecture delivered by Urbain Jean Joseph Leverrier, codiscoverer of the planet Neptune. The talk so captivated Goldschmidt that—with money from the sale of the Galileo portrait—he bought a two-inch telescope and a set of star charts, and promptly found thirteen more asteroids.

As Goldschmidt and others pieced together the orbits of these new sightings, the astronomical community came to realize that almost all asteroids congregate in a torus—a fat, doughnut-shaped tube—between Mars and Jupiter. Within this volume of space, which today is known as the main asteroid belt, the minor planets orbit the Sun like the intertwined turnings on a thick ball of string.

With continued observation in the second half of the century, distinct patterns began to emerge from this congested domain. In 1866, for example, University of Indiana mathematics professor Daniel Kirkwood examined the orbits of the eighty-seven minor planets then known and found three major gaps in the asteroid belt. As Kirkwood informed that year's meeting of the American Association for the Advancement of Science in Buffalo, New York, asteroids seem to avoid certain clearly identifiable spaces, which he defined as mathematical "resonance" points with Jupiter: orbital lanes around the Sun where an object would revolve at a simple fraction of Jupiter's twelve-year-long orbit *(pages 80-83)*.

Kirkwood's calculations showed, for example, that gaps occur where an object would complete one orbit in six years (one-half resonance with Jupiter), one orbit in four years (one-third resonance), and one orbit in three years (one-fourth resonance). As Kirkwood saw it, an asteroid falling into a one-half resonance orbit with Jupiter would be tugged by the giant planet at the same point in every second orbit until the intense and repeated pull made the orbit more elliptical, thereby terminating its resonance. The number of empty orbits identified in the asteroid belt—called Kirkwood gaps to honor their finder—now totals eight, and Kirkwood's basic theory about Jupiter's influence remains in force.

SKY STREAKERS

Renegades from Jupiter's gravitational dominion would make their presence known in time, but not until new search methods had enabled observers to detect asteroids in unexpectedly distant and eccentric orbits. The first such technique was long-exposure photography, and the first astronomer to employ it was Maximilian Wolf, director of the Königstuhl Observatory in Heidelberg, Germany. In 1891, Wolf began to survey the asteroid belt with a camera-equipped telescope, an innovation that would reveal a bonanza of new asteroids.

Wolf's method hinged on the capture of starlight by two photographic plates: He exposed the first for an hour, then the second for an hour, and finally reexposed the first for one additional hour. (A clockwork motor kept the star images in tight focus by turning the telescope and its attached camera in time with Earth's rotation.) An asteroid, because it moves with respect to the background stars, would show up on the first plate as a line with a gap in it, and on the second plate as a line that fitted neatly inside the gap on the first. This two-plate system kept Wolf from interpreting random cracks in the emulsion on one of the glass plates as new asteroids.

No longer constrained to observe the heavens with one eye glued to a telescope, astronomers needed only to distinguish between the pinpoint of light made by a star, the streak of light left by a minor planet tracking across the sky, and the odd scratch caused by careless handling. In forty-one years of studying the photographic images so produced, Wolf discovered a grand total of 231 asteroids.

Long-exposure astrophotography was such a dramatic advance over previous observing methods that a number of astronomers adopted it before the turn of the century. Among the new practitioners of the craft were Gustav Witt, director of Berlin's Urania Observatory, and his assistant Felix Linke.

On the night of August 13, 1898, Witt was using his camera to search for

A long streak of light marks the course of the first observed Aten asteroid on this time-exposure photograph, taken by discoverer Eleanor Helin in 1976 with the Palomar Observatory's Schmidt telescope. A week after the asteroid was detected, it passed Earth at a distance of 11.3 million miles.

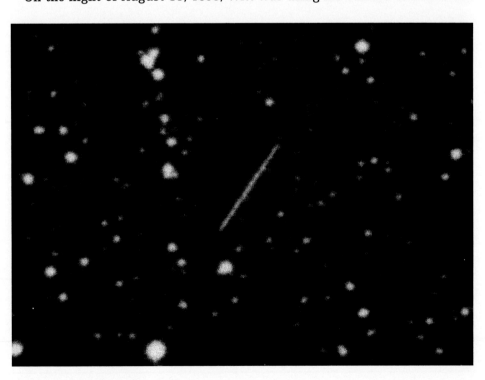

Denizens of the Transition Zone

From studies of asteroids' orbital periods, astronomers have deduced that most of these minor planets inhabit a transitional region between the rocky inner planets and the gaseous outer ones—the so-called main asteroid belt. Following slightly elliptical orbits, these chunks of rock travel in the same direction as the planets and take from three to six years to complete a full circuit. (Although the asteroids' orbits describe a giant torus around the ecliptic, or plane of Earth's orbit, the belt is represented above as a flat disk for clarity.) Two densely populated asteroid clusters—the Greeks and the Trojans, whose component asteroids are individually named for heroes from Homer's *Iliad*—orbit well beyond the main belt, adhering to the same path as Jupiter.

In 1866, astronomer Daniel Kirkwood discovered that certain orbital periods were missing from the main belt. That is, no asteroids could be located in orbits at certain distances from the Sun. These vacant orbits, called Kirkwood gaps, represent regions where Jupiter's powerful gravitational attraction disrupts orbital dynamics enough to kick asteroids into more elliptical paths *(pages 79-87)*, which eventually makes them susceptible to more extreme perturbations. Three categories of asteroids—the Atens, the Apollos, and the Amors—follow highly elliptical and often inclined orbits that cross the path of Mars and sometimes that of Earth *(box, right)*. For example, Icarus, a member of the Apollo group, journeys inside the track of Mercury to reach perihelion at only 18 million miles from the Sun.

Other asteroids, such as 250-mile-wide Chiron, follow tracks that are far removed from the main belt. Chiron's path brings it no closer than the orbit of Saturn on its inbound leg before the asteroid heads back out to the neighborhood of Uranus.

wide main asteroid belt is home to millions of rocky bodies, about 5,000 of them larger than nine miles in diameter. Beyond the main belt, two clusters march in lockstep with Jupiter: The Greeks precede the planet in its orbit around the Sun; the Trojans follow. Unlike asteroids that were perturbed into inclined and elliptical orbits by Jupiter *(below),* the two platoons are able to maintain their positions because they oscillate around so-called Lagrangian points—locations in space where the gravity of the giant planet is balanced by the gravitational tugs of other bodies in the Solar System.

EARTH APPROACHERS

In this sampling of Earth approachers—asteroids whose eccentric orbits bring them close to the third planet—the asteroid named Ra-Shalom *(1)* represents the Atens, a class of asteroids each of whose aphelion, or greatest orbital distance from the Sun, lies just outside Earth's orbit *(blue).* The Apollo class of asteroids is represented by Geographos *(2),* which at perihelion, or closest approach to the Sun, swings inside Earth's orbit and at aphelion ventures close to Mars. Typifying the third class of Earth approachers—the Amors—is the asteroid Eros *(3),* which crosses Mars's orbit *(red)* but does not quite reach Earth's. Astronomers have determined the orbits of at least 140 of these Earth-approaching asteroids so far.

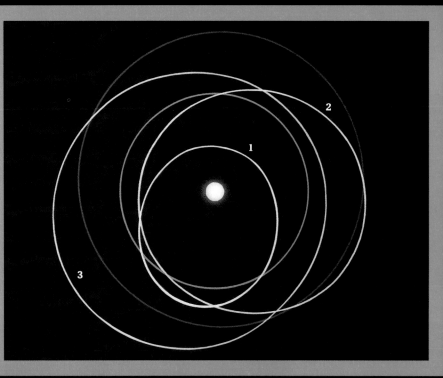

ORBITAL BERTHS IN THE ASTEROID BELT

The asteroids thronging the Solar System may be classed in roughly a dozen categories based on chemical composition, color, and reflectivity. Scientists have found a strong correlation between a given body's relative position and its composition: The farther from the Sun, the less dense the asteroid.

Most fall into just three groups. Nearest the Sun, and accounting for roughly 15 percent of the belt's popu-

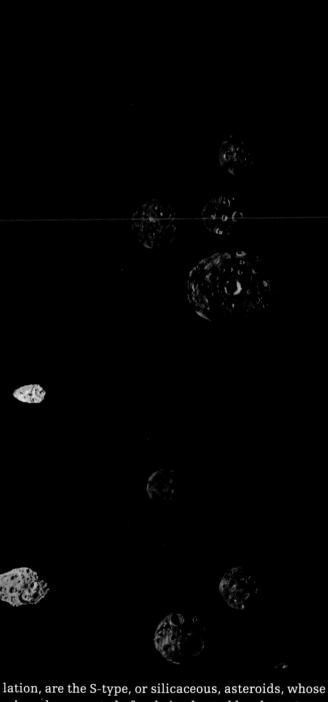

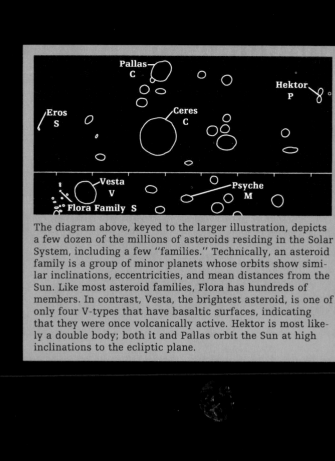

The diagram above, keyed to the larger illustration, depicts a few dozen of the millions of asteroids residing in the Solar System, including a few "families." Technically, an asteroid family is a group of minor planets whose orbits show similar inclinations, eccentricities, and mean distances from the Sun. Like most asteroid families, Flora has hundreds of members. In contrast, Vesta, the brightest asteroid, is one of only four V-types that have basaltic surfaces, indicating that they were once volcanically active. Hektor is most likely a double body; both it and Pallas orbit the Sun at high inclinations to the ecliptic plane.

lation, are the S-type, or silicaceous, asteroids, whose minerals separated after being heated by electromagnetic currents blowing outward from the rapidly spinning young Sun. These most likely consist of olivine around a core of nickel-iron alloy.

Silver-colored M-types—metallic chunks of nickel and iron that emerged from repeated collisions among S-types and others—dominate the belt's middle region

and constitute slightly less than 10 percent of the asteroid family. Carbonaceous, or carbon-rich, C-type asteroids make up about three-quarters of the belt population. Inhabiting the belt's outer regions, the C-types were never heated to the melting point; they have thus preserved much of their original chemical makeup, giving astronomers a window into the formation of the Solar System 4.6 billion years ago.

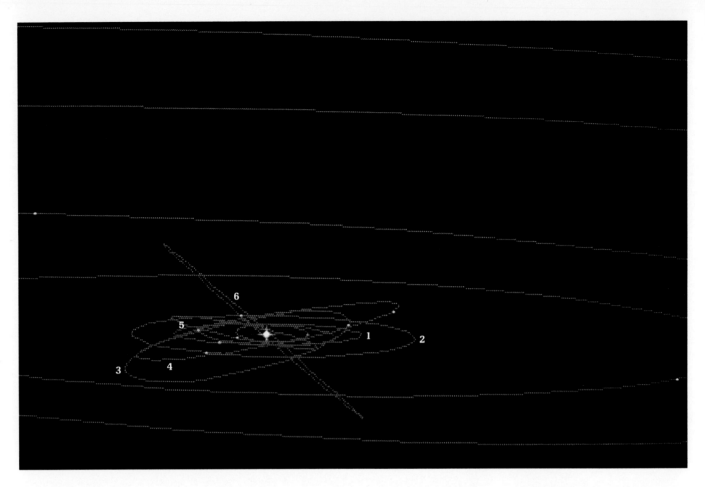

the asteroid Eunike, which had been spotted twenty years earlier but whose orbit had never been precisely determined. Witt found not only Eunike and another known asteroid but something else as well. As he and Linke examined their photographic plates on the morning of the fourteenth, a third line on the plate captured the scientists' attention: "It was so long that we first thought it to be a flaw in the emulsion," Linke recalled. "But it was too clean to be that, hence we suspected an actual object with a high rate of apparent motion, a comet."

That evening, after taking a closer look at the object through the observatory's twelve-inch refracting telescope, Witt and Linke concluded that it could not be a comet, because it lacked a tail. It therefore had to be an asteroid—and, as indicated by the length of the streak on the two photographic plates, it had to be traveling at uncommon speed. A laborious calculation of the asteroid's orbit, completed by Witt's colleague Adolf Berberich thirteen days later, showed that it was indeed unique; the object followed a path so highly elliptical that it ventured far inside the main asteroid belt, even crossing the orbit of Mars on its approach to the Sun. To underscore its novelty, the asteroid was named after a god—Eros—rather than a goddess.

Eight years later, another renegade turned up on a pair of Maximilian Wolf's photographic plates. In contrast to Eros, this asteroid was traveling beyond the outer extreme of the main belt; it appeared to orbit the Sun in the same path as Jupiter, preceding the giant planet by a constant angle of 60 degrees. The oddity was christened Achilles.

The close orbital companionship between Jupiter and Achilles flummoxed

This computer-generated map shows the path, relative to the ecliptic plane, of four asteroids known to cross the orbits of Earth *(1)* or Mars *(2)*. Three of the asteroids charted—Geographos *(3)*, Eros *(4)*, and Ra-Shalom *(5)*—are typical of the minor planets in that their orbits are only slightly inclined. The fourth, Lick *(6)*, revolves at an unusually steep angle—perhaps, astronomers theorize, because the asteroid is a defunct comet.

the astronomers who examined it. To avoid being obliterated by Jupiter, Achilles would have to travel at precisely the same speed as the planet—a most improbable arrangement, since the slightest change in the rate of either body would eventually cause the two of them to collide.

Scientists puzzled over this phenomenon throughout 1906. Finally, Carl Vilhelm Ludvig Charlier of Sweden's Lund Observatory recalled a 134-year-old analysis that solved the mystery of Achilles' survival: In 1772, Charlier reminded his colleagues, French mathematician Joseph-Louis Lagrange had written a visionary paper, "Essay on the Problem of Three Bodies," in which he suggested that gravitational interactions between the Sun and a planet could create a stable zone for a third (and smaller) celestial object at the exact point where Achilles had just been found. This balancing act could only occur, said Lagrange, when the three bodies—Sun, planet, and planetary minion—formed the corners of an equilateral triangle: Such a configuration would create two pockets of gravitational stability, one of them at a point 60 degrees in advance of the planet and the other at a point 60 degrees behind it.

Lagrange's eighteenth-century musings served to instruct as well as inspire astronomers of the twentieth century. Realizing that Achilles occupies a gravitational safe haven precisely where Lagrange had said one should exist, observers began to examine the analogous point 60 degrees behind Jupiter. Sure enough, by the end of 1906 they had found an asteroid orbiting there too. In honor of their propounder, the two loci, and their equivalents along the orbital paths of other planets, were named Lagrangian points. (They are also called libration points, from the Latin word for "balancing.")

Since then, nearly 200 asteroids have been discovered near Jupiter's Lagrangian points, with roughly two-thirds of them clustered near the forward point. Astronomers have estimated that as many as 2,000 asteroids larger than ten miles in diameter—that is, about half the number of main-belt asteroids of comparable size—may be concentrated around the Lagrangian points. To distinguish the two groups, those in the lead are named for Greek warriors, while the laggards are named for Trojan warriors. The exceptions are Hektor, a Trojan asteroid that infiltrated the Greek camp before the nomenclature was formalized, and Patroclus, a Greek among the Trojans. No asteroids have been spotted in the Lagrangian points for Earth or Mars, but current theories of the Solar System's makeup do not exclude the possibility that such planetary attendants will someday be found.

FRACTURED FAMILIES

Having traced the broad outlines of the main asteroid belt, astronomers set out to divine its subtler patterns. It soon became apparent that asteroids are a gregarious breed of celestial objects, clumping together in groups with as many as 200 members.

Thanks to his extensive writings about the phenomenon, Japanese astronomer Kiyotsugu Hirayama is most often credited with the discovery of these asteroidal clans. But Daniel Kirkwood had been aware of them three decades before Hirayama published his first work. In 1888, Kirkwood pointed out four pairs of asteroids that displayed twinlike similarities: Not only did their orbits incline at nearly identical angles to the ecliptic plane, he noted, but their orbital eccentricities and their distances from the Sun matched as well. Later that year, Irish astronomer William Henry Stanley Monck learned of Kirkwood's results and proceeded to identify an additional eight pairs. "The resemblance frequently extends beyond a single pair," wrote Monck, "and embraces what may be called a family."

Although Monck's label was apt, it did not gain astronomical currency until Hirayama published his first paper, in 1918, describing what he too called "families" of asteroids. In Hirayama's scheme, each family consisted of the fragments of a larger, parent body—but not necessarily a full-fledged planet, as Heinrich Olbers had suggested—that had shattered long ago, perhaps by colliding with another minor planet. He named three large families after the

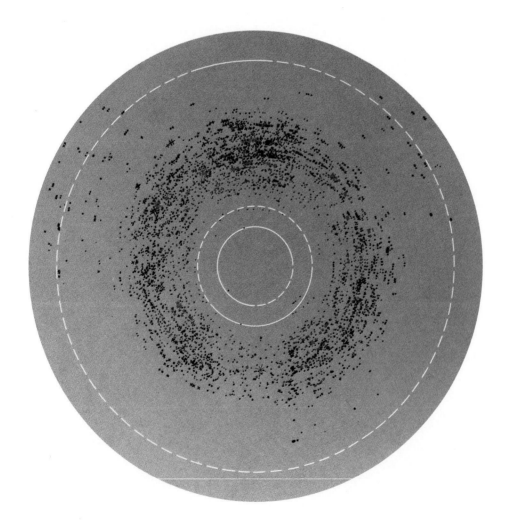

In the mid-1980s, scientists used images made by the Infrared Astronomical Satellite (IRAS) to create this asteroid map, which is based on more than 7,000 sightings of 1,811 known asteroids. The sightings were plotted with reference to the orbits of Earth (innermost ring), Mars, and Jupiter. The map distinguishes between cooler asteroids (red) such as S-types, which reflect the Sun's rays, and warmer ones (black) such as C-types, which emerge clearly only in this type of heat-sensing infrared survey.

largest asteroid in each group: According to one recent count, the Themis family has about 220 members, the Eos family approximately 200, and the Koronis family about 140. Hirayama added Maria and Flora in 1923 and another five in 1928.

Studies of the color and speed of individual family members, as well as the mass distribution of the family as a whole, have supported and expanded Hirayama's theory about the stormy origins of the groupings. In the immediate aftermath of these long-ago collisions, astronomers suppose, each member of a family continued to follow an orbit that passed through the spot where the original smash-up occurred. Over time, however, the complex effects of a host of gravitational interactions caused these orbits to shift, so that some family members now share only roughly similar eccentricities, inclinations, and distances from the Sun.

Recent observations and calculations have suggested that one-third to one-half of the nearly 4,500 asteroids that have been cataloged to date travel

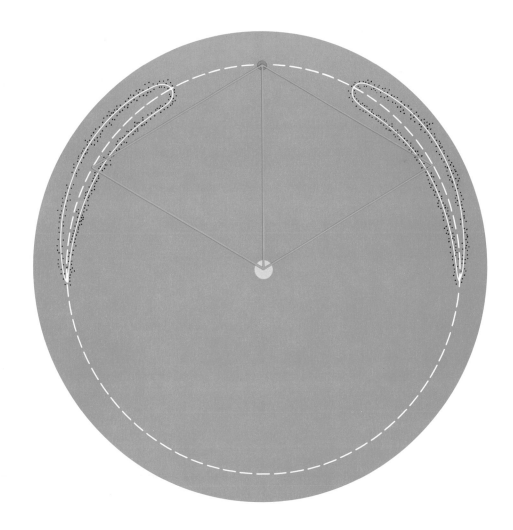

The chart at right details an intriguing feature of the *IRAS* map: the two groups of darker asteroids *(black dots)* located near Jupiter's orbit on either side of the planet. Known respectively as Greeks (the forerunners) and Trojans (the followers), each group centers around a Lagrangian point, the vertex of an equilateral triangle whose base extends from the Sun to Jupiter. Like tethered captives, the asteroids stray only so far from this point of gravitational equilibrium— and avoid colliding with Jupiter.

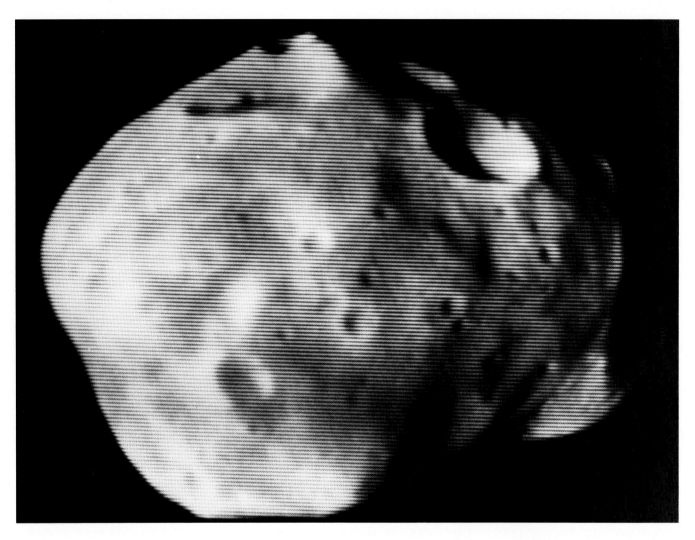

In 1989, a Soviet space probe transmitted this photograph of the cratered terrain of the Martian moon Phobos, thought to be a former asteroid that was caught in the planet's gravitational field. Like many asteroids, Phobos bears the scars of innumerable collisions with other rocky bodies. The encounter that formed the six-mile-wide Stickney crater *(upper right)* was so violent that Phobos nearly shattered; lesser impacts left smaller craters on the moon's surface *(top)*.

in 200 or so families, continually battering one another. For these minor planets—as, indeed, for all asteroids—violence is a way of life: The smaller family members are slowly fragmenting their larger siblings to bits.

AN ASTEROID WITH EARTH'S NAME ON IT?

Dramatic as collisions between asteroids must surely be, the run-in between two barren rocks in space can hardly compare with the destruction that would result if a sizable asteroid were to strike Earth. According to Charles Kowal, an astronomer at the Space Telescope Science Institute in Baltimore, Maryland, the explosion caused by an asteroid the size of 1989 FC hitting the third planet at thousands of miles per hour would dwarf the detonation of all the world's nuclear arsenals.

As long ago as 1932, Belgian astronomer Eugène Delporte discovered an asteroid likely to leave just such a lasting impression on Earth. He spotted a small object circling the Sun on a path whose perihelion lies just 10 million miles from Earth. Delporte named the asteroid Amor, who in ancient Roman mythology was the son of Venus and either Mercury or Mars.

A month after Delporte's discovery, German astronomer Karl Reinmuth unwittingly inaugurated a decades-long competition to sight close-passing asteroids when he spied one that swooped even nearer. Named Apollo, after the Greek god of light, the asteroid actually crossed Earth's orbit just seven million miles behind the planet. Amor and Apollo became the prototypes and namesakes of two distinct classes of minor planets, collectively known as "Earth approachers" or "near-Earth asteroids."

In 1936, Delporte found an Apollo asteroid he called Adonis, which passed by Earth at a distance of 1.25 million miles. A year after that, Reinmuth sighted the Apollo asteroid Hermes, which came within 500,000 miles—the closest brush until the arrival of another Apollo, Henry Holt's 1989 FC, which passed 100,000 miles closer. A little less than forty years later, planetary scientist Eleanor Helin of the Jet Propulsion Laboratory (JPL) in Pasadena, California, would discover a third class of Earth-approaching asteroids, the Atens, whose average orbital distances from the Sun lie inside Earth's orbit.

As astronomers have learned more and more about asteroids, their home planet has seemed less and less safe. By 1990, they had cataloged sixty-five Amors, sixty-three Apollos, and nine Atens—a total that may be but the tip of the Earth-approaching iceberg. Extrapolating from sky surveys and computations of asteroid orbits that he has made over the last two decades, astrogeologist Eugene Shoemaker of the U.S. Geological Survey has estimated that nearly 1,000 additional Earth approachers larger than half a mile in diameter—plus innumerable still-smaller ones—remain to be discovered. Because the Holt asteroid returns about once a year, it poses the most likely danger to Earth.

Lunar and terrestrial cratering records suggest that an estimated fifteen Earth-approaching asteroids crash into Earth or its satellite every million years. Because the rate appears to have remained fairly constant over the last three billion years or so, astronomers looked for mechanisms that would replenish the population of Earth approachers. They came up with two likely sources of resupply. Some Earth approachers, they believe, started out in stable orbits in the main asteroid belt but were launched on Earth-crossing paths by a combination of collisions and the gravitational perturbations of Jupiter or Mars. Other Earth approachers began their celestial lives as short-period comets, then became collision-course asteroids once their gases had been completely evaporated by the Sun.

The orbital and physical characteristics of at least five asteroids lend credence to the second mechanism. One asteroid likely to have once been a comet is 1984 BC, which crosses the paths of Mars and Jupiter on a trajectory strongly reminiscent of those typical of short-period comets. Exceptionally dark, even for an asteroid, with a 2.5-mile diameter similar to that of most cometary nuclei, 1984 BC may well be a comet nucleus whose dust and gases have streamed away during repeated loops around the Sun.

Though intriguing, the origins of the Earth approachers are of less immediate significance than their paths—especially should those journeys end somewhere on the third planet. To detect the approach of these potential Earth crushers, JPL's Helin and the Geological Survey's Shoemaker began a celestial vigil, known as the Palomar Planet-Crossing Asteroid Survey, in 1973. Their technique required a human observer to scan a pair of matching photographic plates with a stereo microscope in search of an asteroid's telltale streak of light, and it led to Helin's discovery of Aten in January 1976. The technique is time-consuming, however, and in 1983 astronomer Tom Gehrels unveiled an automatic detection device known as the Spacewatch Telescope at Kitt Peak Observatory near Tucson, Arizona.

The assemblage features a thirty-six-inch optical telescope with an attached charge-coupled device, or CCD, which converts the light reflected by an asteroid into digitized brightness values that can be processed by a computer. On each night of an observing run, the motor-driven telescope repeatedly scans three regions of the sky. The computer compares the images captured early in the evening with those made later on and flags any unexpected patterns of movement for analysis. (To avoid false alarms, familiar moving objects such as known asteroids and comets are prerecorded in the computer's memory.) Gehrels's telescope is the prototype of a new generation of telescopic cameras, armed with CCDs, that may someday render nonautomated photographic surveys obsolete.

AN ARSENAL OF NEW TECHNIQUES
As scientists improved their knowledge of asteroid orbits in the 1970s and 1980s, they sought to make similar advances in understanding the sizes, shapes, and compositions of the minor planets. Their work progressed on several parallel tracks, with some breakthroughs stemming from new instruments and others resulting from new insights.

The most basic sizing technique, occultation, is actually an old method updated with fancy electronics. It involves timing the passage of a celestial body—in this case, an asteroid—in front of a star. Having timed how many seconds the star blinked out, and knowing how fast the asteroid is traveling, astronomers can calculate the asteroid's diameter. The path of the asteroid's shadow across Earth, called its occultation track, can be precisely predicted only a few days or hours in advance, so the method is usually practiced by amateur observers able to rush highly portable equipment into the field on short notice. The technique's renaissance has been fueled by the development of inexpensive electronic light sensors (used to detect fluctuations in starlight), high-speed digital recording devices, and good amateur telescopes.

Another established observing tool that has delivered new glimpses of the asteroids is photometry, or the science of measuring light—specifically, changes in the intensity of light reflected or emitted by a celestial object over time. Because asteroids are irregularly shaped and spin about their axes, they reflect alternating amounts of light to Earth; this variance manifests itself as a pattern of rising and falling brightness values known as a light curve.

In 1949, astronomer Gerard Kuiper, then at the McDonald Observatory in

Fort Davis, Texas, began a systematic survey of asteroid light curves using a photometer coupled to a telescope; the device registered fluctuations in the light intensity of an asteroid over a period of several hours. Kuiper's measurements became the cornerstone of a massive data bank, containing precise light curves and spin rates for more than 500 asteroids, that has been compiled over the last four decades by astronomers at the University of Arizona, at JPL in California, and at observatories in Sweden, Italy, and many other parts of the world. Because the shape of an asteroid's light curve often reveals the shape of the asteroid itself, the data bank has helped to establish that a number of Earth approachers are more elongated than most main-belt asteroids. Further analysis has demonstrated that an asteroid's average rotational period is nine hours; the extremes are Icarus, which completes a rotation in two and a half hours, and Elsa, which requires more than three days.

During the 1970s, the advent of spectrophotometry allowed astronomers to determine the composition of asteroids by comparing the brightness of the light they reflect in the visible, ultraviolet, and infrared portions of the spectrum with that emitted by the Sun. In 1975, Clark Chapman of the Planetary Science Institute in Tucson, Arizona, designed an asteroid classification scheme predicated on the fact that various minerals reflect the Sun's light in distinctive ways. Silicates, for example, are known to reflect visible light mostly at the red end of the spectrum, so asteroids that look reddish at optical wavelengths are primarily silicaceous (classified as S-types). By the same token, asteroids that look black or dark gray in the optical range are either carbonaceous (C-types) or metallic (M-types). To distinguish between the latter two groups, an astronomer uses satellite observations to examine their reflectance spectra in the ultraviolet range, where the C-types reflect less UV light.

An asteroid that reflects relatively little visible light will often reveal its vital statistics in the infrared range. The heat radiated in this part of the spectrum can disclose an asteroid's direction of rotation, its overall shape, and—with 90 percent accuracy—its diameter. However, carbon dioxide and water vapor in Earth's atmosphere screen out many of the longer wavelengths of infrared energy from space, hindering the ability of ground-based observatories to study such emissions.

In August 1989, an Earth-approaching asteroid named 1989 PB was captured in a sequence of radar images, shown here and on the next three pages. Obtained over a few hours at a range of 3.5 million miles by the Arecibo radio telescope, the images show the double-lobed mile-wide asteroid making slightly less than one rotation on its axis. Such double asteroids may be produced when two chunks of similar size, moving slowly relative to each other, gently gravitate together—a notable exception to the rule of violent minor-planet encounters.

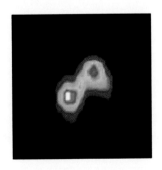 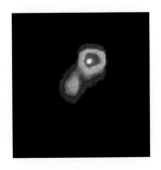 

Astronomers therefore welcomed the 1983 launch of NASA's Infrared Astronomical Satellite, designed to scan the heavens for signals in the far infrared. In ten months of observations, *IRAS* recorded the infrared emissions from 1,811 known asteroids. The satellite also delivered some unexpected news: Two large concentric dust bands orbit the Sun at 10 degrees above and parallel to the ecliptic plane, while a second pair of dust bands orbit in roughly the same position below the ecliptic. Astronomers have tentatively identified the bands as debris from repeated collisions among members of the Eos, Themis, and Koronis families.

RADAR PORTRAITS

An even more sophisticated recent addition to the observer's repertoire—one that has yielded a treasure trove of data on the contours, topographies, compositions, and densities of dozens of asteroids—is radar, the technique of bouncing focused radio waves off distant objects. Because radar signals fade rapidly with distance, the technique has proved most useful for studying Earth-approaching objects.

In mid-1989, astronomer Steven Ostro of JPL used radio waves to capture an image of a paired asteroid tumbling by Earth at a distance of about 3.5 million miles. Dubbed 1989 PB upon its discovery by Eleanor Helin on August 9 of that year, the body turned out to be about one mile long, spinning end over end every four hours, its two portions evidently held together by their own weak gravity. Upon learning of the object's approach from Helin, Ostro rushed to the Arecibo radio observatory in Puerto Rico and trained the facility's radar beam on the path of the object in time to catch a quick look. The fused asteroids circle the Sun every 400 days, an orbital period whose asynchronization with Earth's 365-day orbit dictates that the asteroid twins come within radar range of the planet only at fifty-year intervals.

This expanded arsenal of observing techniques made it so much easier to detect asteroids that during the period from 1970 to 1988 the rate of discovery climbed to an average of 700 per year. Since 1947, keeping track of these myriad bodies has been the responsibility of the Minor Planet Center, which also serves as an information clearinghouse—fleshing out the orbits of newly spotted asteroids, for example—and as an arbiter in disputes over naming and classification.

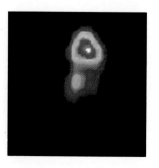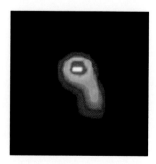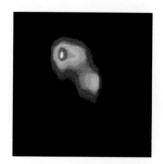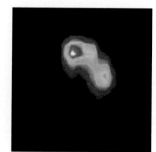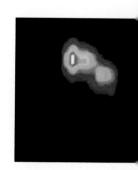

The nomenclature of asteroids has been troubled ever since Piazzi failed in his bid to append "Ferdinandea" to Ceres in 1801. By the late twentieth century, as hundreds upon hundreds of new asteroids were sighted, all restraint had vanished. Minor planets were named after flowers (Azalea, Tulipa, Petunia), wives, wines, public figures (Karl Marx, Herberta and Hooveria, Evita), writers (Chekhov, Tolkien, Mark Twain), composers (Bach, Beethoven, Tchaikovsky), a computer (the NORC), a space-faring television character (Mr. Spock), rock stars (John, Paul, George, and Ringo), and even an airline (Swissair). Although the flippancy has rankled traditionalists, the Minor Planet Center has upheld the principle that discoverers may choose whatever name they like, so long as it is not obscene and the asteroid's orbit has been well established in three passings (two for Apollos).

Classifying the asteroids has proved almost as difficult as keeping their names in order. The difficulty stems from the asteroids' sheer diversity: When classified according to their composition or spectral reflectance, the minor planets fall into at least fourteen separate categories. The C-types are the most abundant, making up 75 percent of the objects in the main belt. The light they reflect indicates that their surfaces consist of carbon-rich clay that has undergone "aqueous alteration"—astronomers' argot meaning that water seeped through the rocky minerals at some point in their formation, partly decomposing them into clay. The brighter, reddish S-types, which may contain traces of iron in addition to their predominant silicates, are the next most plentiful. M-types are rare in number and dull in color, and the scarcest of all are the V-types—bright Vesta is one of the group's four known members—which are unusual in that they apparently consist of basaltic rock, an indicator of molten lava flows.

PLANETESIMAL SOUP

One of the by-products of compositional studies, despite the profusion of categories, is that scientists have gained insights into the asteroids' probable origins. Largely discredited as a result of recent work in this area was German astronomer Heinrich Olbers's notion that asteroids are the fragments of an ancient, planet-size explosion. As late as the 1970s, the idea retained a smattering of adherents, who cited an unproved theory dealing with gravity's requirements for the distribution of mass in the Solar System to argue that

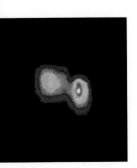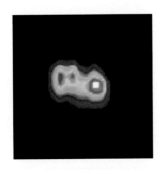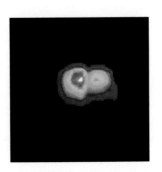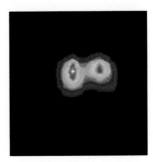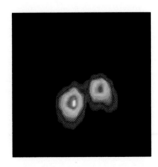

a large planet once must have orbited where the asteroid belt now exists.

However, the highly ordered distribution of asteroid types in the main belt strongly militates against this hypothesis. From its middle to its outer rim, the asteroid belt is dominated by black, carbonaceous rocks; its center contains most of the dull, metallic objects, and its inner rim is loaded with bright silicate S-types. As Clark Chapman has pointed out, the debris of a planetary cataclysm would most likely be scattered about in highly inclined, highly elliptical paths intersecting near the site of the supposed explosion, rather than orbiting the Sun in the orderly, roughly circular, unidirectional paths that most asteroids are known to follow.

Theorists who discount the exploding planet have proposed an alternative means of asteroid creation. During the birth of the Solar System, most of the clumps of matter rotating in a thin, flat disk around the proto-Sun accreted to form the nine planets. Others succumbed to gravitational perturbations and were ejected from the Solar System. A third batch of these planetesimals, or infinitesimal planets, became asteroids. The planetesimals near Jupiter's current orbit were too Sun warmed to preserve their gases; they did, however, keep certain primordial materials, notably water ice and carbon, and evolved over millennia into the C-type asteroids. Those nearest Earth—the present-day S-type asteroids—were even warmer; they lost any ices and other volatile components they may have once had, and bared shiny silicate faces to the universe.

Traces of molten material detected in the few Vesta-type asteroids suggest that some asteroids were heated to the melting point. One theory ascribes the heating to the spontaneous decay of radioactive aluminum in the asteroids' interiors; according to this view, the aluminum isotope was supplied by the same supernova that precipitated the collapse of the solar nebula. A second theory credits the heating to the turbulence of the young Sun, which may once have rotated at fifty times its current speed and spewed forth the streams of charged particles known as the solar wind at speeds and concentrations many times stronger than those recorded today. Such conditions would have bathed the asteroids with electromagnetic fields of an intensity sufficient to induce electrical currents that melted them from within.

Another mysterious facet of the asteroids' evolution is that they have apparently accelerated in their orbits over time. Early in the life of the Solar

 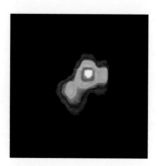 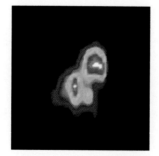

System, the planetesimals that formed planets must have orbited the Sun in nearly identical, circular paths, with negligible velocity differences between neighboring objects. Had their speed relative to one another been much faster than half a kilometer per second, they never would have agglomerated into the larger planets. By contrast, asteroids today travel in rather oblong, tilted orbits, and their average relative velocity is five kilometers per second. Whatever sped them up probably acted while the planets were forming.

Most astronomers cite Jupiter, with a mass 320 times that of Earth, as the accelerating force. Others have suggested that an Earth-size rogue planet might have wandered into the main asteroid belt and churned through the planetesimals before merging with Jupiter or being ousted from the Solar System by the gravitational slingshot forces of that planetary giant. No evidence has yet been found to confirm or deny either one of these theories—but future research may settle the question.

What is certain is that the asteroids' high speed has prevented them from uniting into a planet, consigning them to what Chapman has characterized as a "demolition-derby" existence. Ultimately, they should disappear entirely, crashing into planets such as Mars or Jupiter, being shattered by repeated collisions into undetectable fragments and dust, or suffering expulsion from the Solar System altogether.

MOTHER LODES IN SPACE

If current plans for exploring the Solar System come to pass, researchers will soon get a chance to see asteroids up close. The next advance may come from the Hubble Space Telescope, launched into orbit on April 24, 1990. Although its main mirror is of medium size—less than eight feet in diameter—the Hubble will enjoy a huge advantage over Earth-based telescopes. Free of the planet's obscuring atmosphere, the satellite will be able to distinguish, for example, whether any asteroids besides the pair observed by JPL's Steven Ostro are traveling in tandem.

Along with NASA's shuttle-borne Space Infrared Telescope Facility, or SIRTF, a device scheduled for launch at century's end, the Hubble will capture the entire spectrum of light reflected by an asteroid's surface minerals. With "true" spectral images—so called because their colors and emissions will not be distorted by Earth's atmosphere—astronomers should garner a more precise determination of the asteroids' chemical makeup.

Asteroid watchers will zoom in even closer to their subjects when the interplanetary spacecraft Galileo—launched in 1989—flies within about 1,000 miles of the asteroids Gaspra in 1991 and Ida in 1993. These are representative S-type asteroids in the main belt, elongated in shape and roughly ten to twenty miles across, but their surfaces will be the first ones ever photographed from short range.

For all their ingenuity, none of these remote-observation methods will be as exciting as actually setting foot on an asteroid, which some enthusiasts believe humans will do soon. In 1986, former NASA director Thomas Paine predicted that "the economic development of the Moon, the asteroids, and Mars is going to be a main investment opportunity of the next century." A minor-planet mission might even pay for itself: A solid nickel-iron asteroid half a mile in diameter, mineralogists have calculated, could fetch five trillion dollars. Astronomers Jonathan Gradie of the University of Hawaii and Edward Tedesco and Steven Ostro of JPL have identified two asteroids of precisely this type that travel close enough to Earth to be considered potentially accessible ore deposits. In 1986, one of them came within 23 million miles of the planet, and the other passed just 19 million miles away.

According to astronomers and engineers who have analyzed the economics of mining an asteroid, the most cost-effective approach would be to saddle the asteroid with an electromagnetic "mass driver" and herd it home to Earth orbit. The mass driver would use the asteroid itself as fuel, propelling it through the vacuum of space by ejecting pieces of it rearward at high velocities. Although this means of propulsion would consume about a fourth of the asteroid in transit, enough raw material should remain intact to make several terrestrial fortunes.

That space scientists are giving serious thought to such an undertaking demonstrates how far asteroids have progressed in the esteem of astronomers over the last 200 years. From observational curiosities that showed up in place of an expected planet at the dawn of the nineteenth century, they have evolved into potential suppliers of mineral resources for the twenty-first. Their newfound status, astronomer Charles Kowal observed in 1988, explains why asteroids are likely to play a sizable role in humankind's future: "They are treasures, waiting to be seized."

OF CHAOS AND COLLISIONS

When German astronomer Johannes Kepler speculated in 1596 that an undetected planet must lie in the apparently empty reaches between Mars and Jupiter, he sparked investigations that led to the discovery of the first asteroid, 570-mile-wide Ceres, in 1801. Since then, sky watchers looking for clues to the makeup and dynamics of the early Solar System have plotted the orbits of thousands of these rocky bodies and begun investigating their composition.

With recent advances in both theory and instrumentation have come plausible answers to some longstanding puzzles, illustrated on the pages that follow. For example, astronomers using sophisticated computers and a new field of science known as chaos theory suggest that asteroids in certain orbits—corresponding to the empty lanes known as Kirkwood gaps, after their nineteenth-century discoverer—interact gravitationally with Jupiter in a way that is inherently chaotic. Even though the mathematical equations governing the interaction are well understood, the system is so sensitive to minute effects that in the long term, the result—in this case, an asteroid's orbit—becomes unpredictable. Over millions of years, the chaotic zones are swept clean of asteroids that once inhabited them.

Spectrophotometric studies, which measure the light reflected by an asteroid at various wavelengths, have yielded further insights into the character and habits of the members of the main asteroid belt. A small percentage, for instance, show signs of having been heated to the melting point; others appear to have remained unscathed—except for the scars of collision—down through 4.6 billion years. Collisions, in fact, are inescapable: The asteroids examined today are merely fragments of the original bodies, and the fragmenting process continues. As the asteroids dwindle in size, major impacts will lessen, until all but the largest are ground down, some finally to dust.

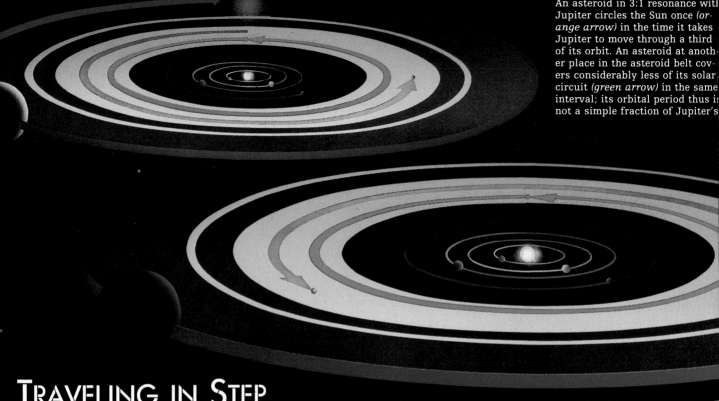

An asteroid in 3:1 resonance with Jupiter circles the Sun once *(orange arrow)* in the time it takes Jupiter to move through a third of its orbit. An asteroid at another place in the asteroid belt covers considerably less of its solar circuit *(green arrow)* in the same interval; its orbital period thus is not a simple fraction of Jupiter's

TRAVELING IN STEP WITH A GIANT

Astronomers believe that as Jupiter accreted from dust and gases some 4.6 billion years ago, its tremendous gravity—about two and a half times that of present-day Earth—sped up the rocky debris lying between it and Mars, thereby preventing the pieces from coming together to form a planet of their own. For a while, some of these bodies were in resonance with Jupiter, circling the Sun at a rate that is a simple fraction of Jupiter's orbital period of twelve Earth years. An asteroid with a four-year orbit, for example, is in 3:1 resonance, making three laps around the Sun in the time it takes Jupiter to complete one. Thus, every twelve years, at its closest approach to Jupiter, the asteroid experiences a gravitational pull strong enough to slightly perturb its orbit. Computer calculations have shown that in the course of millions of years of these repeated tugs, asteroids in certain resonance zones can change orbits radically and almost at random, an indication that the system is chaotic and essentially unpredictable. As asteroids in the affected zones—2:1, 3:1, 4:1, 5:2, and 7:3—are sent fly-

ing, they encounter the gravity of another planet (frequently Mars), which changes their orbits forever. This process gradually empties the zones, producing the Kirkwood gaps *(pages 82-83)*. Asteroids in nonresonant zones, though influenced by Jupiter's gravity, do not always feel the giant planet's tug at the same place and can maintain their original orbits. Surprisingly, other resonant areas are actually stable: A clump of asteroids known as the Hilda group, for example, orbits in 3:2 resonance with Jupiter.

When the resonant asteroid completes its second trip around the Sun, Jupiter is at the two-thirds mark in its own journey. The non-resonant asteroid, which is approximately halfway along on its second orbit, continues to be out of sync with Jupiter.

Jupiter completes the final leg of its orbit just as the resonant asteroid finishes its third circle of the Sun. Thus, on every third orbit, the asteroid is aligned with Jupiter, and experiences strong gravitational tugging. The nonresonant asteroid is also tugged by Jupiter, but at a variety of points along its orbit so that the tugs tend to cancel each other out.

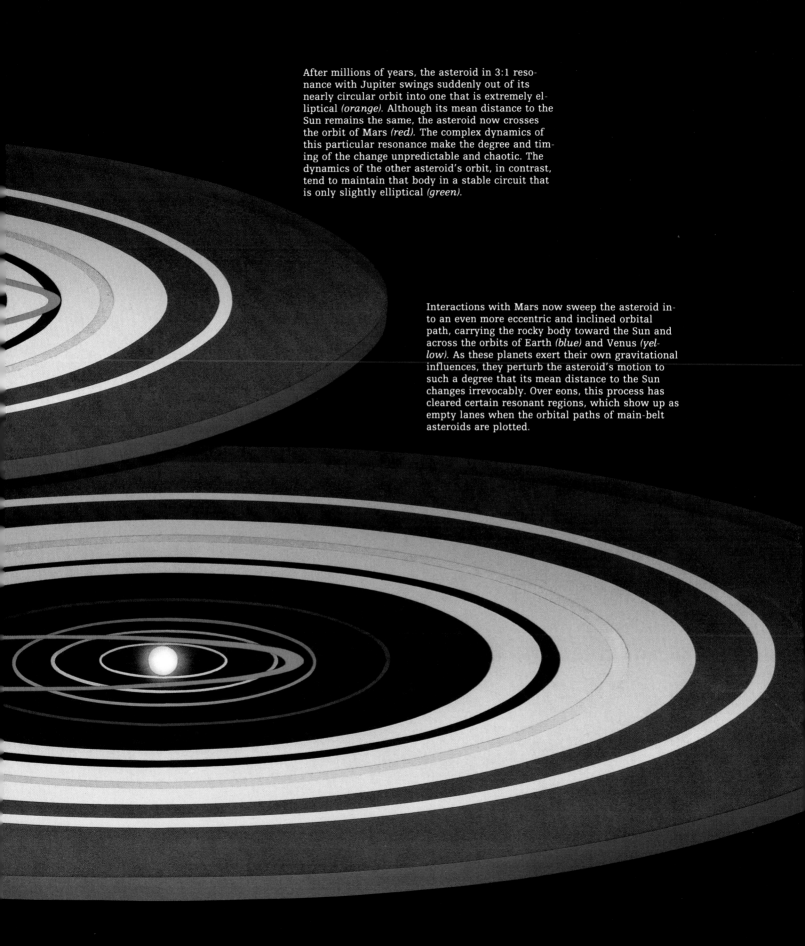

After millions of years, the asteroid in 3:1 resonance with Jupiter swings suddenly out of its nearly circular orbit into one that is extremely elliptical *(orange)*. Although its mean distance to the Sun remains the same, the asteroid now crosses the orbit of Mars *(red)*. The complex dynamics of this particular resonance make the degree and timing of the change unpredictable and chaotic. The dynamics of the other asteroid's orbit, in contrast, tend to maintain that body in a stable circuit that is only slightly elliptical *(green)*.

Interactions with Mars now sweep the asteroid into an even more eccentric and inclined orbital path, carrying the rocky body toward the Sun and across the orbits of Earth *(blue)* and Venus *(yellow)*. As these planets exert their own gravitational influences, they perturb the asteroid's motion to such a degree that its mean distance to the Sun changes irrevocably. Over eons, this process has cleared certain resonant regions, which show up as empty lanes when the orbital paths of main-belt asteroids are plotted.

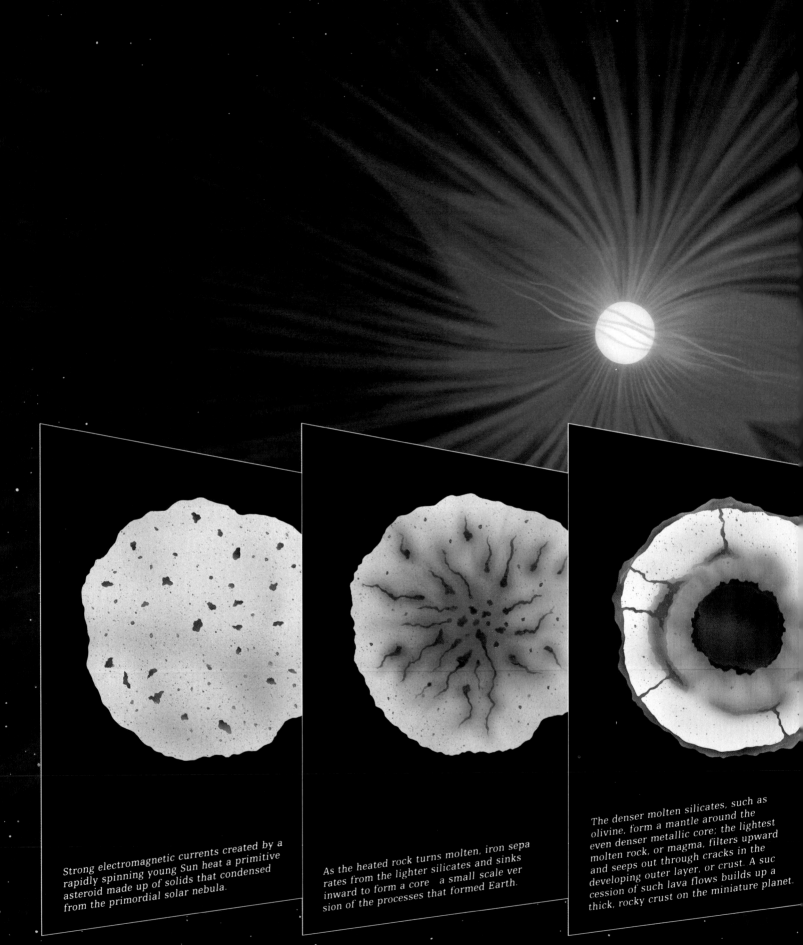

Strong electromagnetic currents created by a rapidly spinning young Sun heat a primitive asteroid made up of solids that condensed from the primordial solar nebula.

As the heated rock turns molten, iron separates from the lighter silicates and sinks inward to form a core a small scale version of the processes that formed Earth.

The denser molten silicates, such as olivine, form a mantle around the even denser metallic core; the lightest molten rock, or magma, filters upward and seeps out through cracks in the developing outer layer, or crust. A succession of such lava flows builds up a thick, rocky crust on the miniature planet.

MELTED BY A SOLAR DYNAMO

Sometime within the first few million years of the Solar System's formation, a number of asteroids in the main belt were heated to the melting point, touching off a segregating process known as differentiation. Heavy minerals such as iron sank to the asteroid's center and formed a dense core. Some silicates, being lighter, percolated to the surface, hardening into a crust. Heavier silicates, rich in olivine, became sandwiched between the crust and the core, forming a region called the mantle.

Scientists have identified two possible sources of heat for this melting. One is the radioactive decay of aluminum-26 isotopes within the asteroids themselves. A second theory, based on recent observations that differentiated asteroids are more abundant closer to the Sun, suggests that the young Sun may be the culprit. Before the Sun evolved into middle age, it rotated much more rapidly than it does today. This rapid spin generated electromagnetic currents that flowed through the asteroids, which then heated up as a result of their natural tendency to resist the currents.

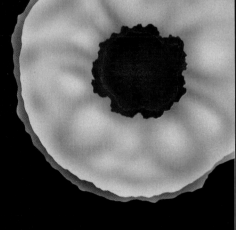

As the proto Sun continues to bathe the evolving asteroid with strong electromagnetic currents, the asteroid's minerals further melt and separate. The resulting high internal temperatures accelerate the differentiation of the asteroid's components.

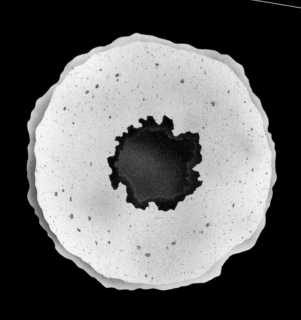

Over the eons, the Sun matures; its spin slows, and its electromagnetic currents subside. As the asteroid cools, its magma hardens into a thick crust. Later collisions may fragment the crust, forming a regolith—a cover of rocky pieces over the bedrock. Finally, olivine crystallizes in the mantle, and the asteroid's iron core solidifies.

THE VIOLENT LIVES OF ASTEROIDS

Like a group of bad insurance risks, virtually all asteroids have been involved in a collision at some point in their lives. By studying the asteroids with photometry—an observational technique in which variations in the intensity of a celestial body's light curve reveal the object's shape, size, and rate of rotation—astronomers have been able to discern the results of these violent encounters.

For all but the very largest asteroids, it turns out, size is no defense against the devastating effects of a collision. A six-mile-wide asteroid orbiting the Sun at three miles per second, for example, carries enough kinetic energy to shatter a neighboring asteroid whose diameter is ten times larger. Spectral examinations of asteroids in the main belt have shown that many such groupings—known to scientists as asteroid families— are the fractured remains of a parent body smashed to pieces by severe impacts.

In the illustration below, the crust of a differentiated asteroid (1) is shattered by a smaller body (2). Another collision shears pieces of mineral-rich crust and mantle from the asteroid (3); further impacts expose the iron core (4). The pieces then orbit the Sun in similar paths as members of an asteroid family (5).

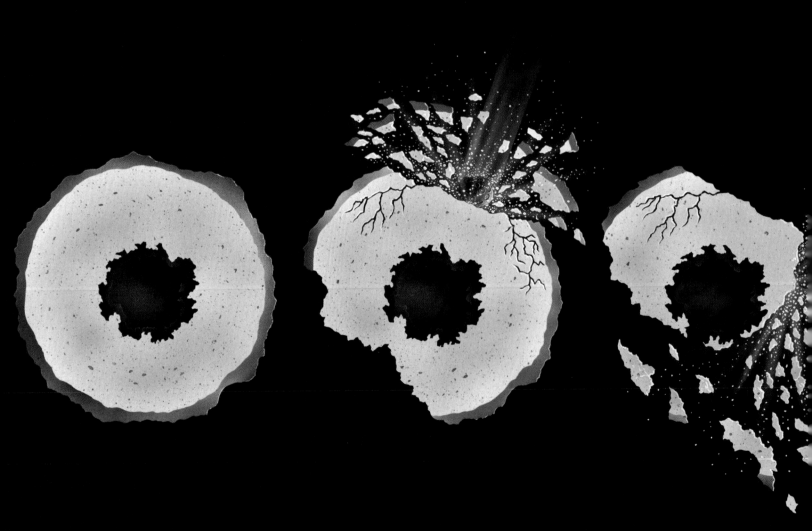

Epochs of repeated collisions have reduced many large asteroids to boulders and particles of dust. As shown above, the process begins when a primitive, or undifferentiated, asteroid *(1)* collides with a smaller body and loses pieces that stay nearby *(2)*. In a collision with a larger object *(3)*, the asteroid is hit with such force that the resulting fragments are hurled beyond the asteroid's own gravitational pull. Smaller impacts, over millions of years, grind the remaining pieces to dust *(4)*.

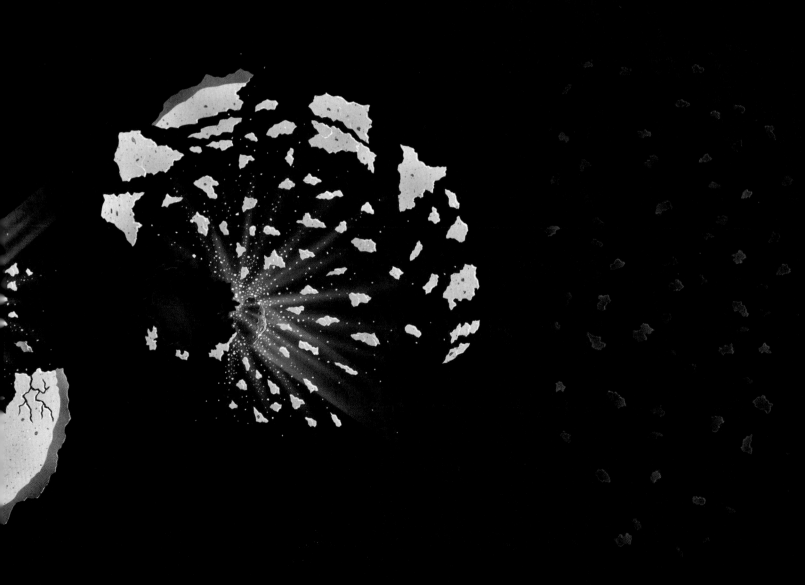

hortly after midnight on February 8, 1969, a dazzling blue-white light shattered the darkness over Pueblito de Allende, a village in northern Mexico. Roused from their beds by a chain of sonic booms, the townspeople watched a brilliant object travel northeast through the sky. It then split in two, and each half detonated in a fiery explosion that spewed out diverging streaks of light. Perhaps an airplane was crashing, the villagers speculated, or a satellite was being incinerated as it reentered Earth's atmosphere. No man-made craft, however, was responsible for this display: It was caused by the midair breakup of a huge rock—a meteoroid—falling to Earth from interplanetary space.

The rock, which may have weighed as much as thirty tons intact, rained fragments by the thousands over the Allende valley below. Word of the fall spread quickly. Within days, geologists, meteorite specialists, and rock hounds from all over North America had arrived to gather the strange stones. The two tons of fragments ultimately collected from the site distinguished the Allende meteorite—Allende for short—as the largest ever observed in the act of reaching the planet's surface. (It would cede that title seven years later, when the residents of Jilin, China, witnessed the fall of a stone whose largest recovered fragment weighs more than all those generated by Allende.)

Allende's arrival could not have been more fortunate for scientists. Not only did it represent a rare and important class of meteorites—the carbonaceous chondrites, whose constituent elements have gone relatively unchanged since the dawn of the Solar System—but its large size furnished a bonanza of samples for various chemical, isotopic, and microscopic analyses. Allende was also recovered promptly, sparing it from excessive terrestrial contamination and weathering. Best of all, it fell just as many laboratories—notably, NASA's Lunar Receiving Laboratory in Houston—were inaugurating new techniques designed to analyze Moon rocks, the first of which would be retrieved by Apollo astronauts in July of that year. The same methods could be adapted to study Allende.

Among the scientists who flocked to the Allende valley was chemist Gerald J. Wasserburg. From his beret and rough field attire, few onlookers could guess that Wasserburg directed a laboratory for the study of lunar and meteoritic samples at the California Institute of Technology in Pasadena. In the next decade, scientists at this facility (fondly known as the Lunatic Asylum) would make a crucial discovery about the Allende meteorite; for the moment, however, Wasserburg was intent on simply finding as many new-fallen specimens as he could.

The fragments were easy to spot. They were dark in color, and many had so-called fusion crusts—hard, glazed layers that formed when the heat of friction with Earth's atmosphere melted the outer part of the meteorite. Beneath this fusion crust was dark gray stone, made up of countless mineral grains interspersed with rounded inclusions called chondrules, which occur in certain meteorites but never in terrestrial rock.

The samples were scattered over an elliptical region that measured about thirty miles long and six miles wide, its long axis parallel to the meteorite's path of travel. The smallest pieces—virtual meteoritic gravel—were found at the incoming end of the ellipse, while the largest piece, weighing as much as fifty pounds, was recovered at the ellipse's far end; in between, the fragments increased steadily in size. This was a textbook case of a so-called strewn field, the natural distribution on the ground of debris from a midair breakup. The gradual increase in size across a strewn field occurs because a small stone has more surface area for its mass than a larger rock of the same shape; it therefore encounters more air resistance than its bigger traveling companions and so falls sooner than they do.

Laboratory tests revealed that Allende dates back to the very beginnings of the Solar System, some 4.6 billion years ago, when it coalesced from the same primitive dust-and-mineral stew that was accreting to form the Sun's nine principal satellites. Like comets and almost all known meteorites, it holds precious clues to the conditions of the Solar System's creation.

TO FIND A FALLING STAR

During the Age of Reason two centuries ago, most scientists vigorously resisted the seemingly folkloric notion that stones could fall from the sky. As the phenomenon was increasingly documented, however, meteoritics—the study of meteors and meteorites—began to flourish. Today, fueled by a bonus crop of specimens that scientists harvested during the 1970s and 1980s, meteoritics is a thriving interdisciplinary pursuit: Geologists, chemists, mineralogists, physicists, and astronomers have all made significant contributions to the emergent field.

The number of specialists involved in tracking down meteorites is rivaled only by the wealth of terms they use to describe their quarry. A meteor is a visual effect—typically, a streak of light glowing fleetingly in the night or daytime sky—that occurs when a solid object from interplanetary space enters Earth's atmosphere at seven to nineteen miles per second (thirty-five to ninety times the speed of sound) and is heated to incandescence by the friction of its passage through the air. Such a projectile, termed a meteoroid, may range in size from a minuscule grain of interplanetary dust that has no appreciable effect on the planet to an asteroid-size body that could wreak death and destruction all over the globe.

Scientists call an exceptionally bright meteor a fireball, while a meteoroid that explodes audibly during its atmospheric transit is known as a bolide (from the Greek word for "missile"). Only when it survives its fiery plunge to the planet's surface is a meteoroid termed a meteorite: Those that are observed in the process of arriving and recovered soon thereafter are called falls, whereas meteorites that fall to the ground unseen but are discovered later on are called finds. Together, the cosmic visitors add anywhere from 100 to 1,000 tons of meteoritic material to the planet each day—a trifling amount in global terms, producing no discernible changes in the overall composition of Earth's crust.

Trajectory analyses have revealed that the distribution of meteorite impacts is relatively uniform all over the world. In rocky terrain like that of the New England states, most recovered meteorites are falls, because a few meteoritic stones lying among the thousands of exposed rocks on a farm usually go unnoticed. (Connecticut, for example, has yet to yield a single find, although one of its towns—Wethersfield—was the target of two falls in just eleven years.) On the Great Plains, by contrast, the topsoil is almost completely free of rocks, so the terrain readily reveals stones from space. Kansas, where an average of one meteorite has turned up in every 1,000 square miles of farmland, boasts a higher meteoritic output than any other state. Worldwide, densely populated countries (notably France, Great Britain, West Germany, and Japan) have exceptional recovery rates, simply because there are more people to find the meteorites.

MYTHS AND MISUNDERSTANDINGS

Humankind's early misunderstanding of meteorites is perhaps best illustrated by the fate of a 280-pound stone that broke through an overcast sky and fell into a wheat field outside the Alsatian village of Ensisheim in 1492. The town's suzerain, who would become Holy Roman Emperor Maximilian I the following year, interpreted the event to mean that his domain now enjoyed divine protection from the encroaching Turks; he therefore directed the villagers to enshrine the object in the local church. Fearing that the talisman might suddenly return to the sky during the night, the townspeople chained it to the church wall.

By the 1700s, popular understanding of meteorites had clarified little about them, and scientists were only serving to muddy the picture. In 1718, for example, cometary pioneer Edmond Halley attributed the appearance of a particularly bright meteor over Europe to the spontaneous combustion of "inflammable sulphereous Vapours" in Earth's atmosphere. Scientific comprehension of the phenomenon hit bottom during 1790, when 300 residents of the French town Barbotan reported the fall of a meteorite, only to have their collective sworn testimony dismissed as mere "folk tales" by France's Royal Academy of Science.

A handful of eighteenth-century scholars did believe that rocks could fall from the sky, but they were almost unanimous in proposing earthly origins

Caltech geochemist Gerald Wasserburg kneels beside a newly discovered fragment of the Allende meteorite, whose shattered pieces rained down on a 180-square-mile area in northern Mexico on February 8, 1969. Laboratory analysis by Wasserburg and others indicated that the meteorite was nearly 4.6 billion years old, most likely a remnant of the primordial materials that formed the Solar System.

for them. Some held that the stones had formed in the atmosphere from vapors rising through volcanoes and crevices in the Earth, while others—French mathematician René Descartes among them—contended that meteorites were airborne particles that had been fused into stones by lightning. A third group put the objects' origins even closer to the ground: When lightning struck rocks on the planet's surface, they contended, it could hurl them high into the air; the rocks then appeared to fall from above when seen from afar. The notion of stormy origins neatly explained the loud rumblings that accompany most meteorite falls; although these are now known to be sonic booms caused by the meteorites' passage through Earth's atmosphere, naturalists of the 1700s interpreted the noise as thunder from distant lightning. Accordingly, meteorites were often called thunderstones.

A surge in the number of reported falls and recovered samples around the close of the eighteenth century began to convince scientists that meteorites were legitimate subjects of investigation. The first such event was a shower of some 200 stones, several of them weighing up to three pounds, that rained down on the medieval city of Siena, Italy, on June 16, 1794. Observers related that a high-altitude cloud had materialized in the evening sky, throwing off smoke, sparks, and flashes of red lightning. (Although smoke and sparks are customary accompaniments of a falling meteoroid, the "flashes of red lightning" are difficult to account for today; they may have been imagined in the collective excitement.) There was then a loud explosion, and stony fragments cascaded down into a meadow and a pond outside the city. A number of English tourists were visiting Siena at the time, and they snapped up most of the collected specimens at high prices; these cosmic mementos were circulated in England upon the travelers' return.

England happened to be the site of a second widely publicized fall, this one in the Yorkshire village of Wold Cottage on December 13, 1795, when a field hand saw a meteorite hit the soil nearby. The stony trophy, weighing fifty-six pounds, was later exhibited in a Gloucester coffeehouse (admission: one shilling) and a London museum.

Yet a third fall occurred in the French village of L'Aigle on April 26, 1803, when more than 3,000 stones clattered down on the rooftops in broad daylight. Among the first scientists to arrive on the scene was French physicist Jean-Baptiste Biot, who had been assigned to investigate the site by the French minister of the interior. In a presentation delivered to the members of the Royal Academy of Science on July 17, Biot reported that all of the fragments collected at L'Aigle were remarkably similar, yet none of them resembled the rock of the surrounding countryside. It was clear to Biot that the thunderstones had fallen from the sky.

This trio of observations coincided with a pair of theoretical breakthroughs. The first advance in theory was made by Ernst Chladni, a German lawyer-turned-physicist whose fascination with the infant science of acoustics had led him to invent a keyboard instrument he called the euphonium. In 1792, Chladni traveled to the University of Göttingen to display the device to his mentor, physics professor Georg Lichtenberg, and the two men began talking about Lichtenberg's recent lectures on meteors and fireballs. Lichtenberg's "main talent," as Chladni later recalled, "was to throw out a few thoughts that gave new perspectives and could lead to further investigations." One of the ideas that was tossed off was Lichtenberg's surmise, based on his observations of the phenomena, that fireballs might be cosmic rather than atmospheric in origin.

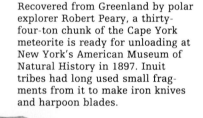

Recovered from Greenland by polar explorer Robert Peary, a thirty-four-ton chunk of the Cape York meteorite is ready for unloading at New York's American Museum of Natural History in 1897. Inuit tribes had long used small fragments from it to make iron knives and harpoon blades.

The notion galvanized Chladni. He sequestered himself in the town's library, where he analyzed the accounts of nineteen meteorite falls that had occurred in recorded history, eight of them since 1700. He also spent time in his laboratory, examining fragments from meteorites made of stone, iron, and a combination of the two. Among the latter was a 1,500-pound stony-iron meteorite that had been recovered near Krasnojarsk, Siberia, in 1749. As Chladni pointed out in a landmark paper published in 1794, no iron ore existed in the region where the meteorite had fallen; a lightning bolt could not have transubstantiated the local rock into iron, he noted, nor could it have lofted a mass of that weight high into the air. To the contention that the stony-iron mass had been smelted by ancient inhabitants of the region, Chladni pointed out that Siberia lacked the iron ore necessary for such a process. Even if the area's primitive peoples had succeeded in such a venture, he added, they could never have blended the stone and iron as uniformly as the two were mixed in the Krasnojarsk meteorite.

Chladni pursued this process of elimination, discounting such proposed origins as lightning, terrestrial vapors, and even the northern lights, to confirm Lichtenberg's suggestion that meteorites come from interplanetary space. They are small masses of matter, Chladni argued, that wander through the heavens until they are captured by Earth's gravitational pull, whereupon they enter the planet's atmosphere and blaze into view as the fireballs that had been documented by Lichtenberg and others. Chladni then hazarded a second and even more farsighted pronouncement, this one about the

The fifteen-ton Willamette iron meteorite, found in 1902, sits on a crude, wooden-wheeled wagon fashioned by the discoverer, Oregon farmer Ellis Hughes (below). Hoping to sell it, Hughes hauled the giant rock three-quarters of a mile to his own property, but a judge ruled that it belonged to the owners of the land on which it fell.

The Hoba meteorite—at sixty-six tons the largest ever found—has never been moved from its landing site in Namibia, where it was discovered in 1920. Had it been much bigger, it probably would have broken apart during its passage through the atmosphere.

origin of meteorites: Either they had failed to join other, larger celestial bodies when the latter were created, he speculated, or they had come into existence when such bodies exploded.

A second major contribution to the acceptance of celestial origins for meteorites was made by English chemist Edward Howard, who performed the first chemical analyses of meteorite specimens. These included two fragments of the Krasnojarsk stony-iron, as well as samples of the Siena stones, some chipped shards from the Wold Cottage fall, and specimens of meteorites recovered in India and present-day Czechoslovakia. Every meteorite examined by Howard—whether a stone, an iron, or a stony-iron—contained nickel, an element that is extremely rare in normal terrestrial rock formations. In bringing to light this common bond among all three types of meteorites, Howard dashed the contention that meteoritic stones might simply be unusual forms of rocks found on Earth.

The work of Chladni and Howard helped meteorites to become accepted as a natural phenomenon deserving of scientific investigation, and nineteenth-century researchers began collecting samples with doggedness and zeal. By 1900, they would amass some 550 stones from space. At first, only rudimentary tools were available for studying and classifying the specimens, but these performed yeoman service. A simple magnifying glass, for example, was

Villagers examine the largest stony meteorite known, a 3,900-pound mass that fell as part of a shower of fragments over Jilin, China, on March 8, 1976.

The three-and-a-half-ton Old Woman, named for the California mountains where it was found in 1976, exhibits the characteristic "thumbprint" markings of an iron meteorite, caused by the flaking off of molten iron during the fiery passage through Earth's atmosphere.

Measuring thirteen feet from end to end and three feet across, this oddly shaped iron meteorite was discovered near the Mexican town of Bacubirito in 1863 by a farmer who struck it with his plow.

used to reveal the presence of chondrules—tiny aggregates of one or more minerals that distinguish stony meteorites. Scientists also employed the technique of displacement—weighing the amount of liquid that a meteorite displaces in a laboratory vessel—to establish that meteorites can be three times denser than typical Earth rocks.

Such crude methods grew in sophistication throughout the 1800s. Chemists, for example, used hydrochloric acid to isolate and identify the element magnesium in the mineral olivine, as well as other basic constituents of stony meteorites. They also used spectroscopy—the technique of analyzing the characteristic wavelengths of light emitted by the elements in a burning powdered sample—to find such additional components as lithium.

COMPLEX CLASSES

With these fairly advanced analytic means at their disposal, the pioneers of meteorite classification were limited not so much by the state of existing technology as by the complex nature of meteorites themselves. Ninety-two percent of meteorites are stones rather than irons or stony-irons, and—as Austrian mineralogist Gustav Rose discovered in 1825—their mineral crystals resist analysis. Rose attempted to study a stony meteorite's mineral grains using a reflecting goniometer, an instrument that identifies unknown minerals by measuring the distinctive angles between the faces of their crystals. To Rose's dismay, the sample crystals were so tiny and so poorly formed that their interfacial angles could not even be gauged, much less compared with those of standard crystal specimens.

Starting in the late 1850s, the introduction of the microscope greatly eased and refined the task of meteorite classification. Austrian chemist Karl von Reichenbach was the first to try to examine meteorite samples using this tool, but he lacked a way to produce sufficiently thin specimens until British mineralogist Henry Sorby hit on a workable technique. Sorby would grind a flat face on a meteorite chip by rubbing the chip against a hard, flat surface coated with the type of mineral abrasives and polishing compounds normally reserved for stone statuary. Using a transparent cement such as Canada balsam, Sorby then glued the face to a glass slide. He employed the same process to grind away the exposed side of the chip until nothing remained but a paper-thin slice. Today this is accomplished by diamond-impregnated saw blades and industrial abrasives.

By illuminating the transparent sheets that resulted from Sorby's method and scrutinizing them under a microscope, scientists could discern the color, structure, and identity of the minerals that were present in each sheet. In 1861, an Oxford professor of mineralogy named Nevil Story-Maskelyne enhanced this method with his introduction of the polarizing microscope. In this device, which is still used today, a light beam is passed through a material that polarizes it, allowing passage only of light waves that vibrate in a certain direction. The polarized light beam is then sent through a meteorite slice, where the beam interacts with the minerals in the sample to pro-

The Meteorite Menagerie

Scientists sort meteorites into three major classes, shown in exterior views and cross sections below. The most common type is the stones, which constitute 92 percent of all meteorites; stony meteorites resemble terrestrial rocks and, like them, are composed mainly

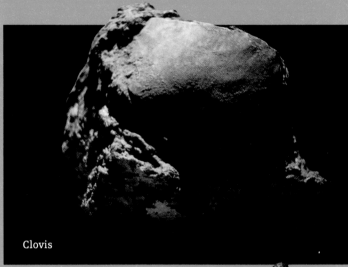

Clovis

Stones. The dome-shaped area of the stony Clovis meteorite, which spent hundreds of years underground before being discovered, may have been sculpted during its fiery passage through the atmosphere. The inside of the Bustee achondrite reveals a conglomeration of rock fragments believed to have been welded together during asteroid collisions that formed it. A larger group of stones, the chondrites, is represented by the Saint Mary's County specimen, seen here in a cross section photographed using a polarizing microscope. The Allende carbonaceous chondrite contains a number of small, dark, irregular grains that may be planetary material preserved relatively unaltered since the birth of the Solar System.

Waingaromia

Irons. Belonging to a common subgroup of the iron meteorites called the octahedrites, the Waingaromia iron *(right)* appears as an unprepossessing remnant of creation when viewed from the outside. Once a fresh surface of such an octahedrite is exposed and etched with nitric acid, however, the meteorite displays an interlocking matrix of nickel-iron alloys known as the Widmannstätten pattern, as seen in the Harriman iron meteorite shown opposite. (The dark circle in the Harriman pattern is an "inclusion," a pattern-breaking clump consisting mainly of troilite, or iron sulfide.) Some irons—the Sikhote-Alin, for example—reveal a much coarser crosshatching, a sign of a lower nickel content.

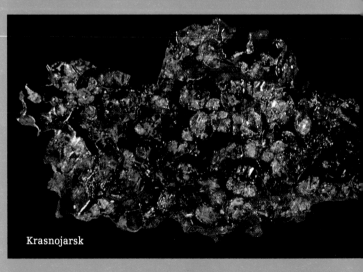

Krasnojarsk

Stony-irons. Three examples of the scarcest class of meteorites, the stony-irons, are shown at right. In the Krasnojarsk meteorite—the first stony-iron ever discovered—nuggets of the mineral olivine sparkle amid the iron-nickel alloy that surrounds them. Stony granules containing the silicates feldspar and pyroxene litter the polished cross section of the Mount Padbury meteorite. The Brenham meteorite features light and dark concentrations of stone and iron; roughly 10 percent of the metal is nickel.

of silicate minerals. The next most numerous group is the irons, which represent seven percent of meteorites and consist mostly of nickel-iron alloys. Stony-irons, the third variety—and the rarest, at only one percent of the total—contain silicates and metals in nearly equal proportions. Of the three classes, the irons dominate the world's meteorite collections because they withstand weathering better than do stones, and because their extraordinary heft makes them easier to recognize as unusual.

Bustee

Saint Mary's County

Allende

Harriman

Sikhote-Alin

Mount Padbury

Brenham

duce distinctive colors and patterns of light and dark stripes. An observer records these properties and compares them with those of known mineral samples in order to identify the meteorite crystals.

Because iron meteorites remain opaque even in very thin slices, Reichenbach also experimented with various chemical methods for displaying their crystalline structure. The most successful of these was a two-stage technique in which he first polished an iron sample, then etched it with acid, revealing surface markings that corresponded to the edges of the crystals within. The most notable feature disclosed by the acid was the Widmannstätten pattern, the term for a striking geometric array *(page 99)* that is found in no earthly iron. (The pattern was named after an Austrian count, Aloys von Widmannstätten, who had been among the first to notice it, in 1808.) Reichenbach's method showed that iron meteorites are made up of three forms of nickel-iron: the minerals kamacite and taenite, as well as a fine-grained intergrowth of the two called plessite.

German physicist Max von Laue added an even more powerful technique to the analytic arsenal in 1912. When an x-ray strikes a crystal lattice, von Laue discovered, the beam is reflected at an angle that depends on the distance between adjacent parallel planes of atoms within the crystal. (Just as the angles between the planes in a crystal are distinctive, so too are the distances.) Whereas the chemical assays of Rose and Reichenbach could identify only which elements were present in a meteorite's minerals, von Laue's method—called x-ray diffraction—revealed the crystal structures of the minerals themselves.

The century-long improvement in classification enabled scientists to divide the three main categories of meteorites into a number of distinct subcategories. Stony meteorites, for example, are classified as chondrites or achondrites, according to whether or not they contain chondrules. Both chondrites and achondrites consist mainly of silicates, crystals whose basic building block is a silicon atom surrounded by four tightly bonded oxygen atoms. Stony-irons, another main category, consist of silicates embedded in nickel-iron; their two major groups are pallasites and mesosiderites. The third major

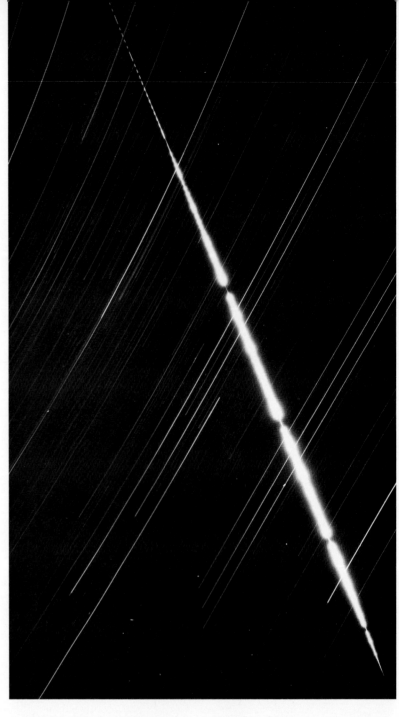

Plunging earthward on the night of January 3, 1970, a meteor appears as a javelin of light in this image captured by an automatic camera, whose shutter action segmented the fiery trail. (The thinner lines are star tracks produced by Earth's rotation.) Computer analysis of the meteor's trajectory led scientists to its impact point near the town of Lost City, Oklahoma, where they recovered a twenty-three-pound, teardrop-shaped stone.

category, the irons, can be divided into at least twenty chemical groups, so scientists find it more convenient to group them according to their structure. The octahedrites, for example, display a detailed Widmannstätten pattern that may reveal their thermal histories—the brief but intense period of heating that the meteorites underwent as they formed long ago.

CRATERS IN THE EARTH

If scientists had been reluctant to accept the extraterrestrial origin of meteorites, they were downright bullheaded when it came to recognizing the geologic scars that such projectiles create as they slam into the planet at 25,000 to 70,000 miles per hour. No matter what the impacting body's angle of incidence, a meteorite that weighs more than about 350 tons will explode upon impact, leaving behind a roughly circular impact crater. Geologists have identified more than 120 such craters all over the globe, some measuring scarcely 500 feet in diameter and others as much as eighty-five miles.

The reality of meteorite craters was in doubt only a century ago. One particularly hot dispute began in 1891, when an East Coast mineralogist

Resembling crater-scarred worlds of their own when photographed under a scanning electron microscope, the minuscule glassy balls known as microtektites *(right)* are the hardened remains of molten silicates flung skyward by meteorite impacts on Earth millions of years ago. The surface markings probably resulted from erosion by seawater, in which the microtektites were suspended.

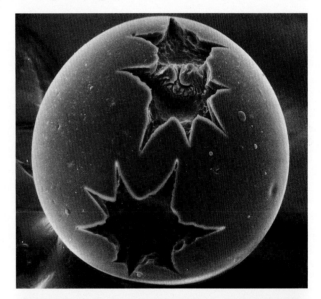

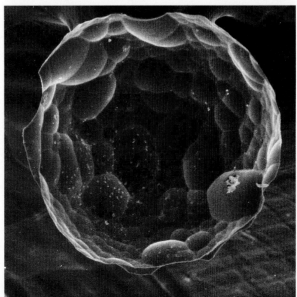

Virtually unaltered by erosion, Arizona's Meteor Crater memorializes an impact that occurred some 50,000 years ago. Scientists estimate that a 300,000-ton amalgam of iron and nickel blasted the pit, which measures more than 4,000 feet across and 600 feet deep. About thirty tons of fragments have been recovered from the area.

named Albert Foote was sent to verify the rumor that a large iron-ore vein lay in the vicinity of a 4,150-foot-wide crater near Winslow, Arizona. Foote examined the iron fragments littering the desert around the crater and decided that they were pieces of a meteorite. He reacted in both scientific and entrepreneurial fashion: After publishing his findings in the *Proceedings of the American Association for the Advancement of Science,* Foote began selling specimens worldwide. Among his customers was Hubert Newton, the Yale College professor who had pointed out the link between comets and meteor showers in 1863; Newton paid Foote $1,250—an astonishing sum at the time—for an iron souvenir weighing 835 pounds.

Foote's work grabbed the attention of Grove Gilbert, the director of the U.S. Geological Survey, who commissioned a more rigorous scrutiny of the crater. Gilbert's survey geologist concurred with the judgment of Foote that the iron was meteoritic in nature; he also reported that the rim of the crater had been peeled backward and that iron fragments had been found as far as eight

METEORITE FINDERS

1794 German physicist Ernst Chladni proposed that meteorites arise in the cosmos, not in Earth's atmosphere.

1802 British chemist Edward Howard performed assays that established the presence in all meteorites of nickel, a metal that is rare on Earth.

1896 After flirting with an impact origin for Meteor Crater, near Winslow, Arizona, American geologist Grove Gilbert rejected the notion for lack of evidence.

miles from the crater—all indications, wrote the geologist, that "a big fellow had made the hole."

The cautious Gilbert teetered on the brink of acceptance. Seeking more conclusive evidence, he argued that if the iron meteorite had buried itself in the ground upon impact, it should betray its presence by causing wild fluctuations in the needle of a magnetic compass held nearby. With this in mind, Gilbert and a survey party visited the crater in 1892, but none of their instruments detected the hypothesized magnetic disturbances. Gilbert therefore embraced an alternative explanation. In 1896, he concluded that the Arizona crater had been formed by a huge underground steam explosion, a form of terrestrial volcanism.

But the matter was far from settled. The most vocal doubter was geologist and local silver-mine owner Daniel Barringer, who learned of the crater in 1902. Barringer rejected Gilbert's steam-explosion theory because no byproducts of volcanic activity, such as sulfur, lava, or radial fissures, were

1923-1972 In half a century of collecting meteorites, self-taught specialist Harvey Nininger used a variety of methods—including the metal detector below—to amass 6,000 specimens.

1902-1929 Arizona prospector-geologist Daniel Barringer proved Gilbert wrong but searched in vain for a solid body beneath Meteor Crater.

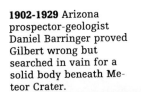

Leveled by a mysterious blast in 1908, trees lie
strewn like matchsticks in the Tunguska region
of Siberia. Meteoritic fragments found at the
site in 1961 suggest that the culprit may have
been a small asteroid or comet that exploded in
midair, devastating almost 800 square miles.

present in the area. "It does not seem possible," he wrote in an attack on Gilbert's reasoning, "that any experienced geologist could have arrived at such a conclusion." Convinced that the crater harbored a metal-rich meteorite, Barringer secured the mining rights to the area and set up his own firm, the Standard Iron Company, to drill into the crater floor. By 1904, the company had extracted five core samples but no large iron mass. However, the test holes yielded trace amounts of diamonds, platinum, and iridium—telltale components of a meteorite—as well as substantial deposits of powdered silica, which Barringer attributed to the disintegration of the local sandstone by the force of a huge impact.

A persuasive argument for the impact theory—and an explanation for Barringer's failure to locate the object that had scooped out the giant crater—was advanced in 1908 by geologist George Merrill of the Smithsonian Institution. Merrill proposed that the Arizona meteorite had broken apart upon impact, producing the many iron fragments found near the crater, and that the intense heat generated by the collision had vaporized most of the rest of the object. Like Barringer, Merrill was struck by the extreme granulation of the powdered silica; this "rock flour," he stated, had been converted from sandstone by "some dynamic agency acting like a sharp and tremendously powerful blow."

As it happened, a stupendous explosion occurred the same year but produced no crater whatsoever. On June 30, a huge meteoroid—perhaps even a small comet—disintegrated in midair over the remote Siberian region of Tunguska with a blast that leveled more than half a million acres of forest *(left)*. Soviet scientists did not reach the site until 1927, and even then they misinterpreted the data they gathered. What they took to be small craters turned out to be normal features of the permafrost terrain, whose flowing subsurface water routinely pockmarks the landscape. Indeed, no meteorites were ever retrieved from the site of the Tunguska event, as modern astronomers have prudently labeled it. However, microscopic particles containing a high concentration of the element iridium—a commonplace constituent of both meteorites and interplanetary dust—would be collected by another Soviet expedition decades later.

Barringer, meanwhile, had dismissed Merrill's results; if true, they meant his quest was pointless. In 1918, he oversaw the drilling of a 1,370-foot-deep hole that turned up a 30-foot-thick layer of nickel-bearing fragments beneath the crater floor. Here at last, Barringer speculated, was the exterior of the mysterious iron mass that he had pursued so long; newly heartened, he financed a string of mining ventures in the crater throughout the 1920s. By the time of his death in 1929, Barringer had sunk $120,000 of his own money into a project that never hit pay dirt.

Barringer's relentless championing of the crater's impact origin did yield dividends of another kind. By the 1930s, most scientists agreed that the crater had been blasted by a meteorite, and the site was renamed Barringer Crater, or Meteor Crater. Corroborating evidence came in the 1960s, when the minerals coesite and stishovite—both produced as the shock wave from a high-speed impacting body passes through rock containing quartz—were found in the local sandstone. Not only that, but excavation beneath the crater floor revealed a huge mass of breccia, or broken rock, which is caused by the instantaneous shattering of bedrock. Mixed through the breccia were millions of tiny metallic spherules, the molten remains of meteoritic vaporization.

A RELENTLESS PURSUER

Around the same time that Daniel Barringer was preparing to dig his final shaft in Arizona's Meteor Crater, a college biology teacher in Kansas witnessed a spectacle that led him to become the world's preeminent collector of meteorites. It was 8:57 on the night of November 9, 1923, and Harvey Nininger was chatting with a fellow professor on the campus of McPherson College. "Suddenly a blazing stream of fire pierced the sky," Nininger recalled half a century later, "lighting the landscape as though Nature had pressed a giant electric switch. The blade of light vanished with equal suddenness, leaving a darkness seeming thicker than before." The meteorite responsible for such a dramatic display, Nininger guessed, must have landed within 150 miles or so. He composed a request for eyewitness accounts of the fall and sent it off to newspapers around the state. The reports he received in return confirmed his estimate: The meteorite had come to Earth near Coldwater, Kansas, just 136 miles from the McPherson campus.

Although the meteorite had brought traffic to a standstill and sent the residents of Coldwater rushing from their homes, no one had recovered a sample. For the next year, therefore, Nininger haunted the Coldwater area, giving talks in schools and prevailing upon local editors to publish additional pleas for information or evidence. "I explained the nature and behavior of meteorites," Nininger recounted, "then offered to pay a good price for any specimen found. I counted on this incentive to alert the whole community to interest in the meteorite." Nininger's strategy worked: Local farmers led him to two rocks—a forty-one-pound oxidized nickel-iron and an eleven-pound stone—that had surfaced during plowing. Neither specimen had actually resulted from the November 9 fall, but Nininger paid the owners a dollar a pound for each one.

Thrilled by the pursuit, Nininger gave up teaching and launched himself into meteorite collecting full-time. His early researches indicated that only fifty-three falls had been recorded in the United States since 1803. Resolving to "modify these statistics considerably," Nininger visited schools and churches throughout the Midwest, delivering hundreds of lectures on how to recognize meteorites and distributing thousands of leaflets encouraging readers to inform him of any discoveries. In 1930, he moved to Denver, Colorado, where for the next thirty years he supported his wife and three children by collecting, classifying, and selling samples to museums, laboratories, and fellow meteoriticists around the world. In 1937, Nininger founded the American Meteorite Laboratory in Denver, which sponsored field searches for meteorites; a decade later, he moved to Arizona and opened the American Meteorite Museum near Meteor Crater. To provision these various enterprises, Nininger traveled to all parts of the globe—including Japan, the Philippines, Vietnam, Australia, and New Zealand—in search of new meteorite falls and finds.

In 1972, after almost fifty years of hunting down stones and metals from space, Nininger estimated that he had distributed more than 200,000 leaflets to the public; in return, he had received 35,000 rocks that people thought might be meteorites. More than 2,000 of those specimens proved to be genuine, and they derived from 222 previously unknown meteorites.

CLUES TO ORIGINS

While meteoriticists like Nininger were working to establish where on Earth meteorites end up, others in the field were attempting to discover where in the Solar System they had begun. Essential to the quest were a pair of sophisticated instruments, the mass spectrometer and the electron microprobe. In the mass spectrometer, introduced in the 1920s, a meteorite sample is vaporized and then ionized—electrified by an electron beam or a spark. The resulting ions, or charged atoms, are then channeled through a magnetic field, which bends their flight paths differentially according to weight: The lighter an ion, the more it is deflected. Separating ions of different atomic masses allows their relative amounts to be measured.

The second tool—the electron microprobe, developed by French physicist Raymond Castaing in 1951—enabled scientists to measure the overall chemical composition of a meteorite. In this instrument, a microscopically thin stream of electrons strikes a polished mineral grain, stimulating the mineral's atomic nuclei, which in turn emit x-rays. Since the wavelengths of x-rays depend on the elements that produced them, a chemist can identify the atoms in a specimen by comparing their x-rays with standard data for the various elements. The intensity of the x-rays of each wavelength reveals the number of atoms of each element in the sample.

With these new analytic tools, scientists overturned the theory, propounded by Chladni back in 1794, that meteorites were spawned by the explosion of a single large planet. The main toppler of the view was University of Chicago chemist Harold Urey, a Nobel laureate who had shifted to planetary science after working on the Manhattan Project during World War II. As Urey pointed out in 1952, an exploding planet would melt from the heat of its breakup, yet many meteorites—notably the carbonaceous chondrites—contain mineral grains whose rough shapes prove that their parent bodies were never molten.

Attempting to divine just what those bodies might be, Urey teamed with fellow University of Chicago chemist Harmon Craig in 1953 to review every existing chemical analysis—286 reports in all—of the largest group of meteorites, the chondrites. Although Urey and Craig were forced to discard 192 of the analyses because they contained flawed or incomplete data, the remaining 94 clearly showed that chondrites fall into two distinct groups: Those high in iron, the H chondrites, contain about 28 percent iron by weight, whereas those low in iron, the more common L chondrites, contain only about 22 percent. Urey and Craig interpreted the division as evidence that chondrites come from at least two different parent bodies, each containing a different amount of iron. Their conclusion was suspect on a number of grounds—for one thing, the H and L chondrites might have formed in regions of differing iron content within a single planet—but in raising the possibility of multiple parent bodies, Urey and Craig had charted the course for decades of future meteorite research.

As it turned out, the parent bodies of meteorites are many indeed. In 1960, three other Chicago chemists—Gordon Goles, Robert Fish, and Edward Anders—calculated that the heat of radioactivity inside a full-size planet would have expelled the planet's internal gases (notably argon, which results from the breakdown of potassium-40) into space. Yet radioactive dating showed that argon had been accumulating in meteoroids since the very birth of the Solar System. Arguing that only a parent body considerably smaller than a planet could have cooled fast enough to retain most of its argon, the Chicagoans proposed that many meteorites once belonged to asteroids no more than a few hundred miles in diameter.

Radioactive dating, the technique on which the Chicago three had based their analysis, is a reliable means of determining a meteorite's age, because radioactive isotopes (unstable forms of the same chemical element) decay at a constant rate. The rate at which a radioisotope decays is expressed as its half-life—that is, the time during which half of the parent atoms are transformed into so-called daughter atoms. Using a mass spectrometer, a scientist can determine the ratio of parent to daughter atoms in a meteorite sample; the scientist then factors in the parent's half-life to compute the sample's age. Radioactive isotopes with exceptionally long half-lives—rubidium-87, for example, whose half-life is an almost unimaginable 49 billion years—have revealed that most meteorites formed at the inception of the Solar System.

So-called cosmogenic isotopes constitute a second type of radioactive clock that has helped astronomers retrace the complex history of meteorites. Produced by exposure to cosmic rays (high-energy subatomic particles that pervade the universe), these isotopes are found in all meteorites. Since most cosmic rays can penetrate stone and metal to a depth of only about three feet, the proportion of cosmogenic isotopes in a meteorite reveals how much time has passed since the object was liberated from its parent body and began basking in the cosmic rays of interplanetary space. Most chondrites appear to have been exposed to cosmic rays for less than 40 million years.

This finding posed a severe challenge to the parentage hypothesized by the three Chicago chemists, appearing to rule out the main asteroid belt as the point of origin for most meteorites: Forty million years is but a fraction of the time required for random gravitational perturbations to dislodge a me-from Shergotty, Nakhla, and Chassigny, the towns where the three major meteorites were recovered). For one thing, argon-dating techniques *(right)* indicated that the SNC samples—including the Antarctic rock below—are only 1.3 billion years old; almost all other meteorites found on Earth are the same age as the Solar System, or about 4.6 billion years old. Also, trace-element studies revealed that all eight stones had undergone several stages of melting and recrystallization, a process that occurs on large, geologically active bodies such as moons and planets.

Astronomers searching for the stones' probable origin quickly ruled out the Moon, whose chemistry is different and where most geological activity is believed to have stopped three billion years ago. They also eliminated Mercury, since nearly all rocks dislodged from that planet would be roped into a tight solar orbit, rarely venturing as far as Earth. Venus was deemed just as unlikely; its thick atmosphere and powerful gravity (88 percent of Earth's) would keep most surface objects from achieving escape velocity. The outer planets, too gaseous to produce solid meteorites, were never serious contenders—which left only Mars. The Red Planet seemed to meet all criteria. First, it is volcanically active, although its volcanoes are dormant at present. Its surface atmosphere is as thin as terrestrial air at fifteen miles high, and its gravity is only 38 percent of Earth's, offering limited resistance to space-bound rock. The most convincing argument, however, was finding the same proportions of certain gases in the Antarctic meteorite as were known to exist in the Martian atmosphere. Having narrowed the list of suspects to one, astronomers now had to figure out how the deed was done. One

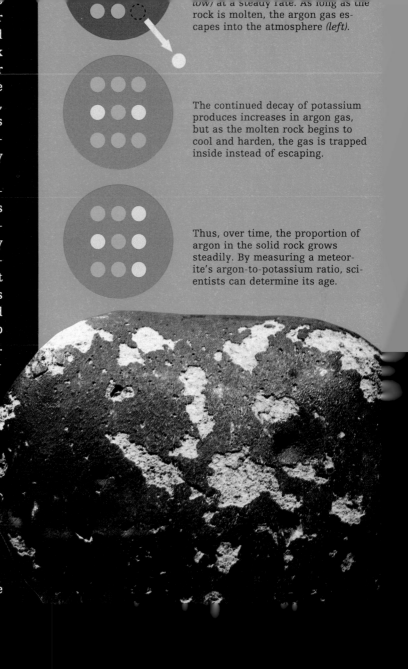

low) at a steady rate. As long as the rock is molten, the argon gas escapes into the atmosphere *(left)*.

The continued decay of potassium produces increases in argon gas, but as the molten rock begins to cool and harden, the gas is trapped inside instead of escaping.

Thus, over time, the proportion of argon in the solid rock grows steadily. By measuring a meteorite's argon-to-potassium ratio, scientists can determine its age.

ESCAPE FROM THE RED PLANET

Despite all the data pointing to Mars as the place of origin for the SNC meteorites, scientists were hard-pressed to explain how rocks that were large enough to have survived the trip to Earth could have been boosted off the Martian surface. The necessary escape velocity of three miles per second, though less than half of that needed to overcome Earth's gravity, is beyond the propulsive force generated by any known type of volcanic activity.

Astronomers considered a second possibility: that the rocks were hurled skyward by the cratering impact of a large asteroid. But this scenario was initially ruled out because it did not jibe with prevailing theory. Any impact large enough to generate shock waves that would accelerate rock to the required speed would, it was thought, crush, melt, and vaporize eve-rything in its vicinity—throwing up nothing bigger than pulverized dust particles.

Then, in the mid-1980s, two University of Arizona scientists, Jay Melosh and Ann Vickery, developed a new theory of crater formation based on painstaking mathematical analysis of the behavior of shock waves near the surface of a target planet. The pair proposed the existence of a relatively shallow zone around the edges of the impact site in which surface rocks would be shielded from the crushing force of the shock waves and instead would be kicked upward at great speed; the larger the impact, the larger the ejected fragments. In fact, the scientists suggested, the shielding at the surface would be so effective that microorganisms in the Martian soil, if they exist, might be hurled unin-jured into space.

When an asteroid crashes into the Martian surface, annihilating itself in a burst of molten and vaporized rock, the impact gouges a crater and sends potent compression waves through the adjacent rock. According to one theory, as the shock waves radiate outward, the most intense pressure is directed down *(white crescent);* rock to either side is subject to less stress. Near the surface, the crushing force of the shock wave is dissipated: Pressed from below and meeting no resistance from overhead, rocks on the surface are heaved upward instead of being melted or crushed to powder.

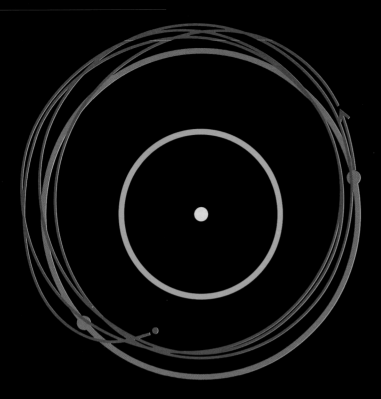

Circling the Sun in elliptical paths *(red)* that cross Mars's orbit *(orange)*, ejected Martian rock spends millions of years shifting randomly in response to the gravity of Jupiter and Mars. Eventually the rock passes close enough to Mars to get a gravitational kick from its parent world. The kick knocks it out of its old path *(below)* into a more eccentric one that could hurl it inward to intersect Earth's orbit *(blue)*.

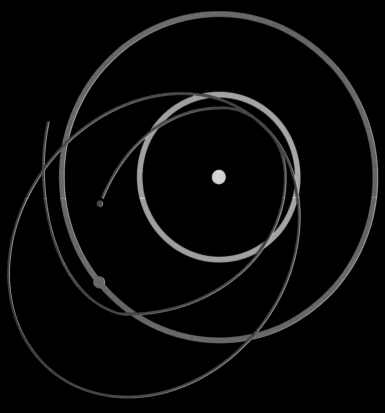

THE LONG JOURNEY TO EARTH

Vickery and Melosh theorized that fragments up to 100 feet in diameter could have been ejected from Mars by asteroids that left crater scars hundreds of miles across. About half the ejected pieces would eventually fall back to the Martian surface, and of the rocks that escaped permanently, about half would eventually be flung out of the Solar System by Jupiter's massive gravity. The remainder would gradually shift orbits as a result of gravitational interactions with Mars until, as illustrated here, some of those orbiting fragments would intersect Earth.

Although the theory seemed to reinforce the Mars-as-origin hypothesis, one discrepancy remained to be accounted for. A technique similar to argon dating indicated that the rocks had suffered shock damage about 200 million years ago, presumably from the event that ejected them from Mars and sent them on the long journey to Earth. But another dating technique, based on nuclear alterations caused by the cosmic-ray bombardment a meteoroid would experience in space, indicated that the SNC samples had been exposed to cosmic rays for no more than 10 million years. The hypothesized solution to the puzzle is described at right.

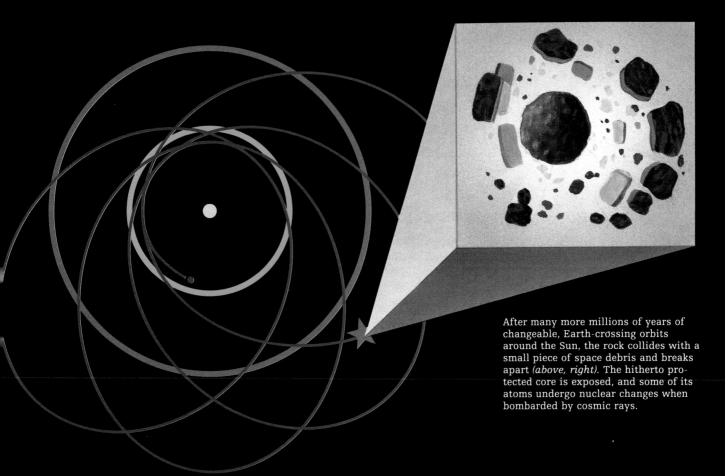

After many more millions of years of changeable, Earth-crossing orbits around the Sun, the rock collides with a small piece of space debris and breaks apart *(above, right)*. The hitherto protected core is exposed, and some of its atoms undergo nuclear changes when bombarded by cosmic rays.

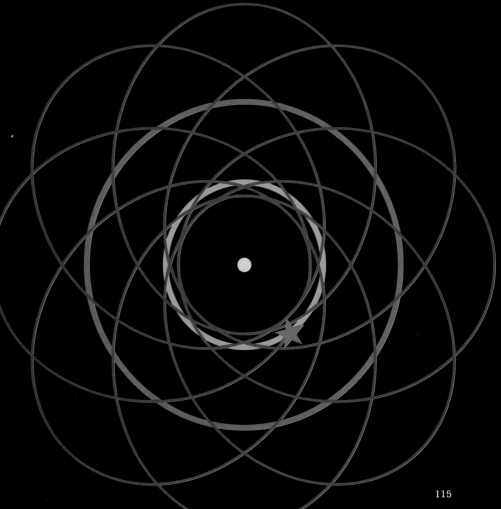

The exposed core continues orbiting the Sun for another 10 million years, following a constantly shifting series of Earth-crossing paths. In time, one orbit carries it so close to Earth that it is captured by the planet's gravity. Plunging through Earth's atmosphere, the rocky scrap comes to rest on the surface of an alien world.

teoroid from its main-belt orbit and set it on a course for Earth. The gravity of Jupiter or collisions between main-belt asteroids could accelerate the process and abbreviate the journey, yet the average estimated transit time between the asteroid belt and the third planet remains in the range of millions of years. By the early 1970s, according to one meteoriticist, "The problem of getting material from belt asteroids to the Earth on time seemed to be insurmountable." But redemption for the Chicago view came in 1982, when Massachusetts Institute of Technology mathematician Jack Wisdom used an elaborate computer model to prove that asteroids falling into a certain Kirkwood gap *(pages 82-83)* in the main belt can be perturbed into new orbits that stretch all the way to Earth in the span of a mere one million years.

By the time of Wisdom's pronouncement, scientists had grown accustomed to startling revelations about meteorites. Six years earlier, for example, Jerry Wasserburg's assemblage of planetary scientists at Caltech's Lunatic Asylum had uncovered evidence that meteorites may bear witness to an event that predates the Solar System's very formation. After exhaustive study of the 1969 Allende meteorite, Wasserburg and fellow geochemists Typhoon Lee and Dimitri Papanastassiou found abnormally high concentrations of the isotope magnesium-26 in a carbonaceous chondrite from Allende. The excess magnesium-26 was identified as the daughter of radioactive aluminum-26, an isotope that may be produced by the massive stellar explosions known as supernovae, among other astrophysical processes. In 1976, Wasserburg's team proposed that Allende's original supply of aluminum-26 had been injected into the solar nebula by a nearby supernova; the concussion from that blast, they suggested, might also have triggered the collapse of the nebula and the subsequent birth of the Solar System itself.

METEORITES ON ICE
One of the most surprising pieces of news about meteoritic origins came from what scientists had believed to be an unlikely spot, Antarctica. Because most meteoroids travel in or near the ecliptic plane, a land surface near either pole lies almost parallel to the paths followed by these solar wanderers; the North and South poles therefore receive fewer falls per square mile than do other regions of the globe. Meteorite specialists thus had long discounted the top and bottom of the world as fertile grounds for meteorite recovery. But in 1969, Japanese glaciologists exploring the Antarctic icecap stumbled upon nine meteorite fragments within an area just a few miles square. Although the proximity of the pieces suggested the strewn field of a single meteoroid, laboratory analysis showed that the nine stones had in fact come from at least four different sources. Because even a single find is surpassingly rare, most astronomers rejected the idea that four unrelated meteoroids could have struck the same small patch of the globe.

A few years later, geologists attacked the puzzle. Some natural terrestrial process, they theorized, must have concentrated the meteorites in the area where they were found. The geologists' view was confirmed by glaciological studies carried out in 1976 and 1977, which reconstructed the mechanics of Antarctica's meteoritic largess: Meteorites that fall on the icecap are covered by snow, which turns to ice as it accumulates; as the ice eventually starts inching downhill to the sea, it carries the trapped meteorites along with it. In certain places near the coast, the drifting ice meets a barrier: a mountain slope or a nunatak (a gentle hill that is actually the summit of a mountain buried in ice). In this way, the rocks that rained down over a huge region are funneled into relatively small catchment areas. Harsh winds blowing across the Antarctic plateau gradually expose the buried stones, as well as cleaning and smoothing the surrounding ice, making its distinctive blue color easy to spot by aerial or satellite surveillance.

Over the last two decades, meteorite hunters mounted on snowmobiles have recovered more than 10,000 Antarctic specimens, which may represent as many as 1,000 separate meteorites. (Until the discovery of these frozen treasure fields, by contrast, just 2,600 meteorites had been collected worldwide.) One of the most fertile grounds is a blue-ice region near the Allan Hills in Victoria Land. On January 18, 1982, American geologist John Schutt discovered a small black stone there that ultimately became the subject of an entire issue of the technical journal *Geophysical Research Letters*. The stone—labeled ALHA81005 because it was the fifth meteorite to be found in the Allan Hills by Expedition A during the austral summer of 1981-1982—merited its star billing for one simple reason: It came from the Moon.

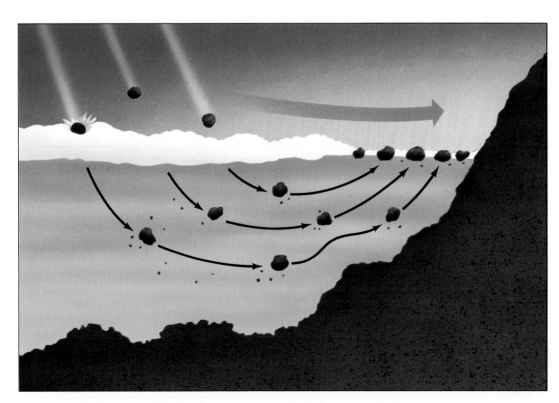

The process of meteorite concentration diagramed at right explains why scientists have collected more than 10,000 specimens in Antarctica. The fallen meteorites become encased in ice, which slowly flows toward the sea. Wherever the ice meets a buried mountain, it is forced upward; fierce Antarctic winds *(blue arrow)* then sublime the ice (that is, they turn it directly from a solid into water vapor), exposing the meteorites and causing them to accumulate on the surface. The patches of wind-swept ice are easy to spot by their distinctive blue color.

At the time, Schutt had no inkling of the stone's lunar origins. That fact emerged only after samples of the meteorite had been sent to two dozen laboratories around the world for analysis. Their tests showed that ALHA81005 contains thousands of small glass spherules and glass-embedded rock particles, both common traits of the Moon rocks gathered by Apollo astronauts from 1969 to 1972. Since 1982, seven additional Antarctic meteorites have turned out to be lunar in origin; planetary scientists believe that these fragments were blasted off the Moon millions of years ago by the impacts of large meteoroids.

Similar collisions may have brought an even more unusual group of meteorites to Earth. All eight stones in the group—called the SNC meteorites, because the first three were found in Shergotty, India; Nakhla, India; and Chassigny, France—are thought to have arisen on Mars *(pages 111-115)*.

A FATAL ENCOUNTER

Of all the recent revelations concerning meteorite origins and destinations, perhaps none has generated more interest than the suggestion that a stone from space wiped out the dinosaurs 65 million years ago. The discovery that led to this theory came in 1978, when physicist Luis Alvarez and his geologist son Walter, working with fellow University of California scientists Frank Asaro and Helen Michel, found abnormally high concentrations of iridium in a layer of underground clay near Gubbio, Italy. Iridium, the platinum-like metal that was found both in Meteor Crater and at the site of the Tunguska event, is scarce in the earth but abundant in asteroids and meteorites. To the Alvarez team, its enrichment at Gubbio strongly suggested a meteorite impact. Most significant, the iridium-filled layer of clay represents the paleontological boundary between the Cretaceous period, when the last dinosaurs roamed Earth, and the Tertiary period, when the age of mammals began. The same meteorite that had deposited the iridium, the researchers concluded, had also done in the dinosaurs.

To test this theory, Luis and Walter Alvarez began to seek out additional evidence for a global catastrophe at the Cretaceous-Tertiary (abbreviated K-T) boundary. After studying K-T boundary cores from fifty sites around the world, the father and son team announced that iridium is indeed overabundant wherever the clay can be sampled. They also stated that the total amount of iridium in the K-T boundary layer, assuming that the element blanketed the

The time line below, based on reported discoveries of iridium-rich clay deposits around the globe, shows the intervals at which scientists believe large meteorites may have hit Earth. Numbers indicate millions of years ago.

590

360

DEATH FROM THE SKIES

Examining the pageant of life encoded in the fossil record of the last 570 million years, scientists have discovered five separate instances in which hundreds of plant and animal species became extinct almost simultaneously. The most intensively studied of these mass extinctions occurred 65 million years ago, when the dinosaurs disappeared from the face of the Earth, bringing the Cretaceous period to a close, and the mammals began their ascendancy, inaugurating the Tertiary period.

In 1980, a team of Berkeley researchers headed by physicist Luis Alvarez and his son, geologist Walter Alvarez, proposed a radical cause for the Cretaceous extinctions: An asteroid approximately six miles in diameter, they argued, had smashed into Earth, throwing as much as 100 trillion tons of dust into the atmosphere. This airborne debris had blocked the Sun's light for months on end, enveloping the planet in a dark, frigid "cosmic winter." As life forms that were sustained by photosynthesis died out, so eventually did the higher orders—notably the dinosaurs—that fed on them.

A wealth of circumstantial evidence, presented below, supports this theory of a cosmic catastrophe. The strongest clue is a thin layer of clay that was deposited worldwide at the end of the Cretaceous period; the clay is exceptionally rich in the metal iridium, which is 10,000 times more abundant in meteorites than it is in Earth's crust.

The quartz grains magnified below support the hypothesis that the dinosaurs perished in the wake of a meteorite impact. The grain on the left is unshocked; the one on the right—from the geologic boundary that marks the end of the Cretaceous period—has shock striations created by the sort of tremendous pressures that stem from a meteorite impact. The boundary-layer clay sample at bottom contains 300 times more iridium than the surrounding rocks. (The coin is about an inch across.)

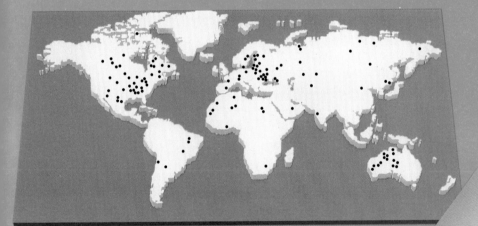

Although meteorites have struck Earth throughout the planet's history, terrestrial weathering and plate tectonics make the scars of these ancient impacts—variously termed craters, impact structures, and astroblemes (Greek for "star wounds")—difficult to identify. Scientists have nonetheless pinpointed more than 100 probable collision sites, whose locations appear above.

160   90   65   38   10   2

entire planet when deposited, would equal the amount contained in a meteorite about six miles in diameter.

An impacting body of that size, Luis Alvarez calculated in 1980, would have blasted a crater ninety miles wide and thrown up enough dust to plunge Earth into darkness for two to twelve months *(pages 121-133)*. With sunlight unable to reach the planet's surface, the photosynthesis that sustains green plants would have come to a halt, and many animals in the food chain—notably those at the top, the dinosaurs—would have died out. Ultimately, the dust raised by the impact would have settled as clay at the bottom of the sea, producing the thin K-T boundary layer observed at Gubbio and elsewhere.

As trees around the world died from this temporary Armageddon, say scientists who specialize in global catastrophes, the desiccated forests would have provided ideal tinder for lightning-ignited fires. Factual support for this scenario was unearthed in 1985, when chemists from the University of Chicago found enough soot particles in the K-T boundary layer to represent the burning of a large portion of all terrestrial vegetation in existence at the end of the Cretaceous period.

Mineralogical evidence, too, supports the impact hypothesis: Stishovite, that unavoidable by-product of meteoritic shock waves, turned up in K-T boundary clay in 1984, as did grains of shocked quartz, which geologists consider to be an equally reliable indicator of a mighty blow. Tiny spheres of sanidine, a mineral forged at the high temperatures typical of celestial collisions, also pervade the boundary layer. Together these disparate clues affirm the pivotal role that the detritus of the Solar System has played—and will continue to play—in shaping the biological history of the planet.

A FATEFUL RENDEZVOUS

Like creatures from another time, some eighty large asteroids periodically venture into Earth's vicinity, dark and swift and traveling at odd angles to the planetary plane. These rocky relics of the Solar System's birth are difficult to spot; scientists estimate their true number to be well over ten times the observed population. The battered face of the Moon testifies to the damage they can do if they collide with another celestial body: Countless craters—many as much as thirty miles across—pock the lunar landscape. From the craters' ages and distribution, planetary geologists have calculated that roughly once every 100 million years an object six miles in diameter hits the Moon. Because the orbits of Earth and the Moon are so closely linked, this impact rate applies to Earth as well. Sooner or later, then, a monstrous asteroid will collide with the third planet.

Projecting what would happen if an asteroid struck Earth has occupied a number of researchers since 1980, when Luis and Walter Alvarez, of the University of California at Berkeley, published a paper linking the extinction of the dinosaurs 65 million years ago with the impact of a projectile several miles wide—a theory that now enjoys broad scientific support. Computer scenarios tracking the short- and long-term events associated with such a calamity recall biblical tales of fire, flood, and famine. One version, illustrated on the following pages, considers the slamming of a six-mile-wide meteorite into the ocean. The energy of the impact—equivalent to the explosion of five billion atom bombs—would transform cool, blue Earth into a flaming crucible. When the smoke cleared, a transmuted planet would emerge: a hobbled and barren world, reeling toward some new destiny.

EARTHBOUND

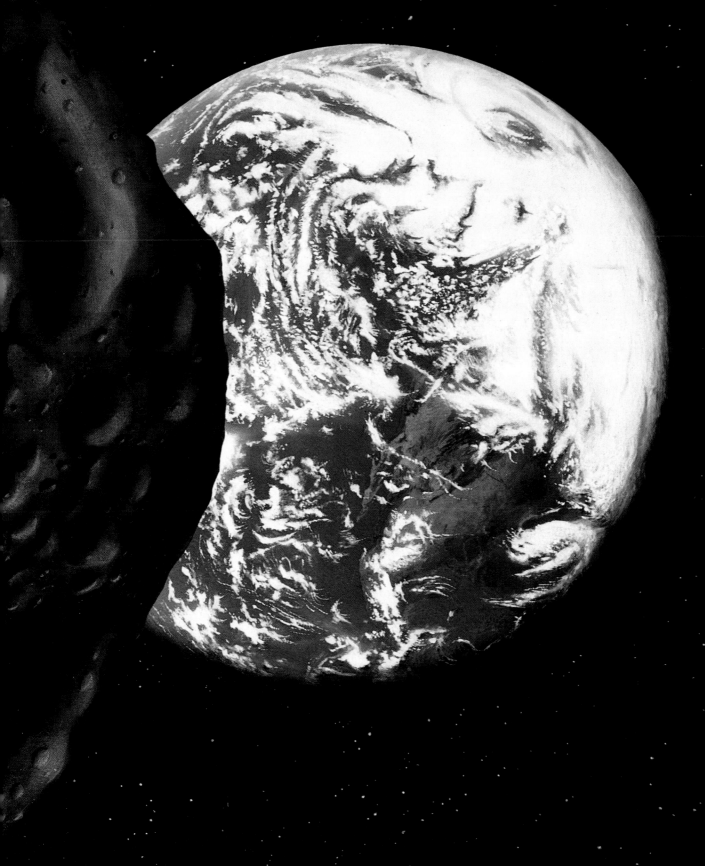

Free Fall

The thin veil of Earth's atmosphere would offer little resistance to a six-mile-wide, trillion-ton asteroid: Diving toward the ocean at 55,000 miles per hour—seventy times the speed of sound—the massive object would blast the air aside, heating the surrounding atmosphere to about 50,000 degrees Fahrenheit. Air molecules, stripped of their electrons by the extreme temperature, would cloak the meteorite in a blazing envelope of visible, ultraviolet, and infrared radiation. Out of this incandescence, a reddish brown smog would materialize: Ionized oxygen and nitrogen would react to form acrid-smelling nitrogen oxide compounds—the progenitors of acid rain. All of this would happen in a fraction of a second.

As the front edge of the huge projectile struck the ocean, impact shocks would instantaneously raise the temperature of the seawater to 100,000 degrees, flash boiling eight trillion tons of brine. Vast jets of vapor would rocket skyward while the asteroid, only three-quarters of a second from seabed impact, would begin a kind of death rattle as rebounding shocks raced through its core.

At the instant of ocean contact, compressed air at the meteorite's forward end is ejected in all directions as jets of white-hot gas, and seawater is compacted to three times its normal density. Reverberating shocks from the ocean-surface impact initiate the breakup of the projectile, though the seafloor, still untouched by the pressure front, remains undisturbed.

IMPACT

When the monster made contact with the ocean bed
100 million megatons of energy would be released
eventually shaking the entire planet. With a stupendous crack of thunder and a blinding flash of light, 100
trillion tons of ocean bedrock and vaporized meteorite
and 130 billion trillion gallons of seawater would
shoot outward from the impact site at 25,000 miles per
hour. In the passage of only three minutes, an expanding fireball of steam and molten ejecta would level any
city within a distance of 1,200 miles and scour the
terrain down to bedrock.

The immense crater produced by the impact, initially 16 miles deep and 200 miles wide, would shrink
to a depth of one-third of a mile and a width of 60 miles
as the underlying mantle rebounded and ocean sediments washed inward. Overhead, shock waves from
the explosion would heat the air to more than 3,000
degrees Fahrenheit, generating searing hurricane-
force winds that would rack the stricken planet for
fifteen to twenty hours.

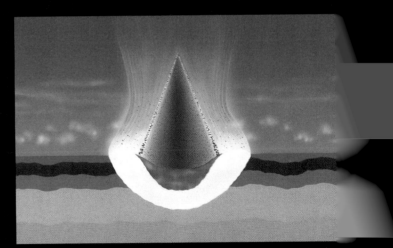

A third of a second after the asteroid first contacts the ocean
surface, shock waves have slammed through the half-mile of water into the seafloor, radiating down into the underlying crust
(light gray) and back up through the water and oncoming projectile. Decompression waves follow, causing rapid expansion of the
compressed water and rock and converting more than 50 percent
of the shock energy into heat. In an explosion equivalent to
250,000 Mount Saint Helens eruptions, trillions of tons of vaporized seabed, meteorite, water, and fragmented rock hurtle out
along the wall of a cavity growing around the point of impact.
Lofted skyward by supersonic airflows, the ejecta quickly fills
the vacuum left by the projectile's passage.

TSUNAMI

The devastating blow delivered by the meteorite would trigger a shift in the seabed 100,000 times more powerful than the earthquake that shook San Francisco in 1906. The cataclysmic ground movement would give rise to a ring of seismic sea waves, or tsunamis, nearly as high as the Rocky Mountains and three to four miles wide.

Were the asteroid to strike in the middle of the Gulf of Mexico *(right)*, the colossal water walls would travel outward at some 450 miles per hour, making landfall simultaneously in New Orleans, Tampa, Havana, and Mérida in the Yucatán. Because of the shallowness of the coastal plains in this region, the waves would roll unchecked as far inland as Kansas City, surge across much of Mexico and Central America, and ravage Florida and the Caribbean islands. Untold numbers of animals would drown, and two million square miles of land would be swamped with silty floodwaters.

Secondary waves spawned by aftershocks and the settling of the crater walls would pound ashore every seven minutes for hours. The relentless churning of the ocean near the impact site would drag warm coastal waters to great depths, suffocating fish and depriving delicate marine organisms of light; frigid bottom waters would inundate reefs and plankton colonies. Decades—perhaps centuries—would pass before all aftershocks ceased and wave patterns returned to their normal state.

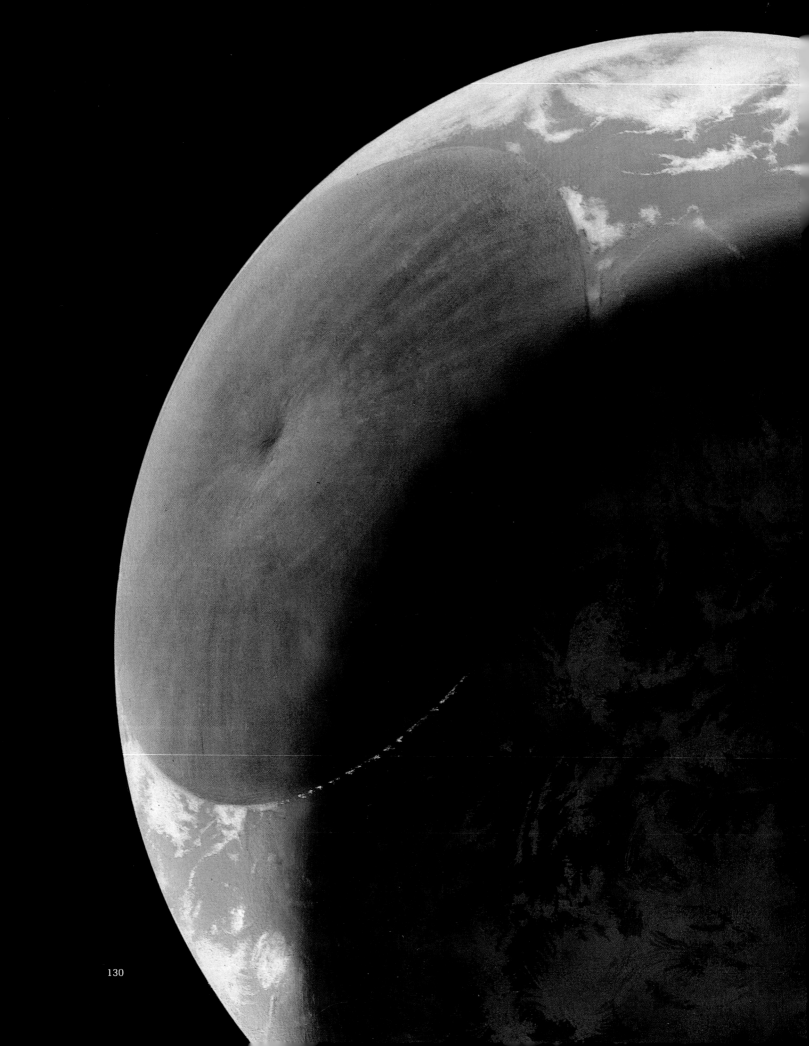

DARKNESS

Like tiny, incendiary ballistic missiles, trillions of tons of microfine rock particles and condensed vapor droplets thrown up by the asteroid impact would soar spaceward, reaching stratospheric heights within seconds. There, in the weatherless realm some thirty miles above the clouds, the fiery grains would rouse violent currents in the cold, thin air.

Propelled by these abnormal winds, the particulates would begin to spread over the planet *(left)*. One theory holds that as the fastest-moving dust grains collided with atmospheric molecules, the resulting friction would impart more energy to each particle than is contained in an equivalent mass of TNT. Ablaze with quadrillions of these glowing bullets, the sky would radiate enough heat to raise the air temperature at an altitude of forty miles to 1,800 degrees. On the ground, temperatures would climb to 600 degrees. Lasting at least an hour and probably longer, the heat pulse would ignite land regions baked dry by the earlier blast winds from the fireball. Creatures unable to shelter themselves from the inferno would be incinerated. Scientists speculate that as much as 90 percent of the world's forests and grasslands untouched by earlier ravages might burn.

Soot from the fires mixed with nitrogen oxide smog produced by the initial and subsequent shock waves would combine with the rapidly spreading dust to form a shroud seventeen miles thick. It would envelop the entire planet within twenty-four hours. Computer simulations indicate that no sunlight could penetrate such a pall; the surface of Earth would be locked away in a blackness thirty times more inky than the darkest moonlit night. For as long as six months, photosynthesis would halt, causing all but the hardiest Arctic plankton to die off and triggering the collapse of the marine food chain. Depending on the season, land plants not already burned would die as well.

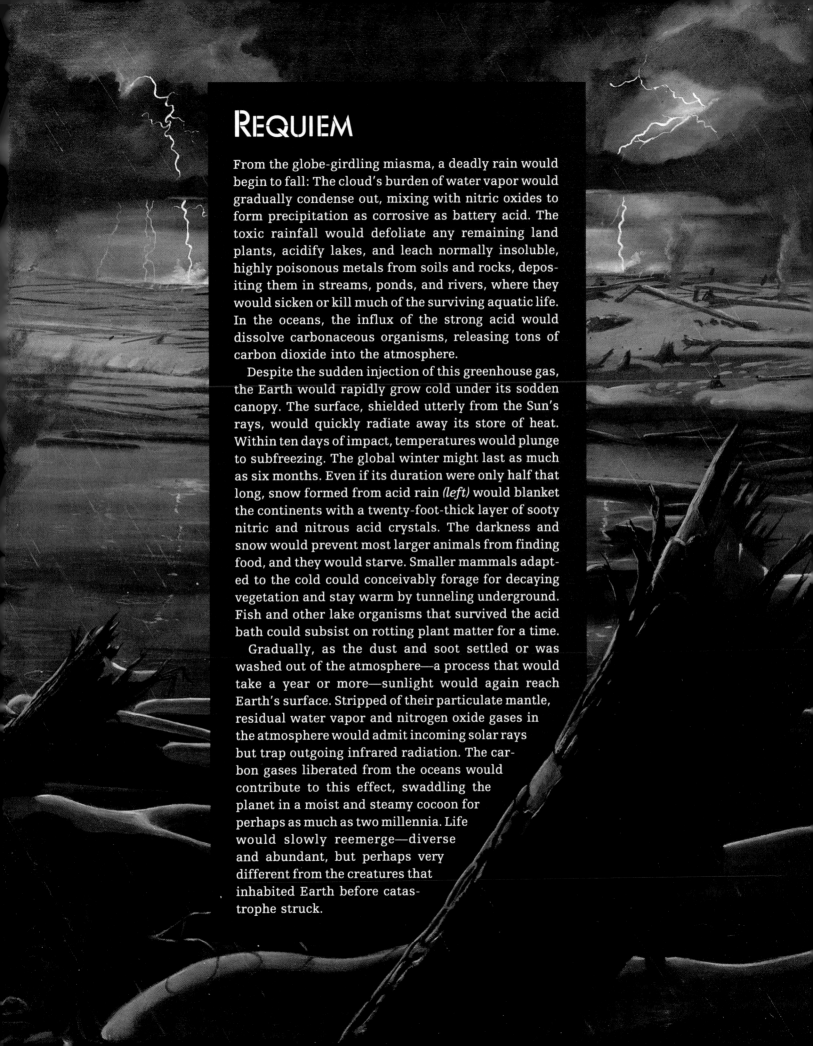

Requiem

From the globe-girdling miasma, a deadly rain would begin to fall: The cloud's burden of water vapor would gradually condense out, mixing with nitric oxides to form precipitation as corrosive as battery acid. The toxic rainfall would defoliate any remaining land plants, acidify lakes, and leach normally insoluble, highly poisonous metals from soils and rocks, depositing them in streams, ponds, and rivers, where they would sicken or kill much of the surviving aquatic life. In the oceans, the influx of the strong acid would dissolve carbonaceous organisms, releasing tons of carbon dioxide into the atmosphere.

Despite the sudden injection of this greenhouse gas, the Earth would rapidly grow cold under its sodden canopy. The surface, shielded utterly from the Sun's rays, would quickly radiate away its store of heat. Within ten days of impact, temperatures would plunge to subfreezing. The global winter might last as much as six months. Even if its duration were only half that long, snow formed from acid rain *(left)* would blanket the continents with a twenty-foot-thick layer of sooty nitric and nitrous acid crystals. The darkness and snow would prevent most larger animals from finding food, and they would starve. Smaller mammals adapted to the cold could conceivably forage for decaying vegetation and stay warm by tunneling underground. Fish and other lake organisms that survived the acid bath could subsist on rotting plant matter for a time.

Gradually, as the dust and soot settled or was washed out of the atmosphere—a process that would take a year or more—sunlight would again reach Earth's surface. Stripped of their particulate mantle, residual water vapor and nitrogen oxide gases in the atmosphere would admit incoming solar rays but trap outgoing infrared radiation. The carbon gases liberated from the oceans would contribute to this effect, swaddling the planet in a moist and steamy cocoon for perhaps as much as two millennia. Life would slowly reemerge—diverse and abundant, but perhaps very different from the creatures that inhabited Earth before catastrophe struck.

GLOSSARY

Accretion: the gradual accumulation of matter in a particular location, usually as a result of gravitational forces.

Achondrite: a stony meteorite characterized by its lack of the tiny, rounded inclusions known as chondrules.

Albedo: the ratio of solar radiation reflected by the surface of an object—be it a planet, satellite, asteroid, or comet—to the solar radiation received by the object; usually expressed as a percentage.

Amor: one of a class of asteroids whose orbits approach, but do not intersect, the orbit of Earth.

Aphelion: the point in the orbit of a celestial object when it is farthest from the Sun.

Apollo: one of a class of asteroids whose orbits intersect that of Earth. The name is also used to designate all three groups of Earth-approaching asteroids: the Amors, the Apollos, and the Atens.

Asteroid: any small, rocky, airless body that orbits a star. Three main kinds have been identified in the Solar System: carbonaceous, or C-type; silicaceous, or S-type; and metallic, or M-type.

Astronomical unit (AU): a unit of measure, often used to express distances within the Solar System, that is equal to the mean distance between Earth and the Sun, or 92,960,116 miles.

Aten: one of a class of asteroids whose average orbital distances are closer to the Sun than is Earth's average orbit.

Basalt: a dark, close-grained igneous rock that is formed by the hardening of lava. Most volcanic rocks are basalts.

Bolide: a meteoroid that explodes audibly during its passage through Earth's atmosphere.

Bow shock: the boundary region of interplanetary space where the solar wind is first deflected by the hydrogen cloud of a comet or by the magnetic field of a planet.

Breccia: rock composed of various rock fragments cemented together by the sudden pressure of an impact.

Calorimeter: a device for measuring heat.

Charge-coupled device (CCD): an electronic array of detectors, usually positioned at a telescope's focus, that registers electromagnetic radiation.

Charge exchange: a subatomic interaction between particles of the solar wind and neutral hydroxyl molecules in the hydrogen cloud surrounding a comet, yielding positively charged ions that become part of the comet's spectacular blue gas tail. *See* Photodissociation; Photoionization.

Chondrite: the most abundant subgroup of stony meteorites, characterized by the presence of chondrules.

Chondrule: a rounded, millimeter-size drop of rapidly cooled silicate, found in abundance in chondritic meteorites. The presence of chondrules distinguishes chondrites from achondrites.

Coesite: a rare mineral that is produced when the shock wave from a meteorite passes through rock containing quartz. It is named after Loring Coes, Jr., the American scientist who originally synthesized the mineral by subjecting quartz to high pressures and temperatures. *See* Stishovite.

Coma: a diffuse region of gas and dust that surrounds the nucleus of a comet and forms the comet's visible head.

Convection: the transfer of heat in a liquid or a gas by the movement of currents from hotter to cooler regions.

Differentiation: a geologic process in which the material of a primitive celestial body such as a planet or an asteroid separates, or is stratified, into regions of differing chemical composition. Differentiation typically produces a core, a mantle, and a crust.

Diffusion: the gradual mixing of the molecules of two or more substances by the movement and scattering of atoms in an element's solid state.

Eccentricity: the degree to which an elliptical orbit deviates from a perfect circle.

Ecliptic: the plane described by Earth's orbit around the Sun.

Electron: a negatively charged particle that normally orbits an atom's nucleus but may also exist independently of it.

Ellipse: a closed, symmetrical curve drawn so that the sum of the distances from any point on the curve to each of two fixed points (called foci) is constant. A circle is an ellipse with its two foci superimposed at the center. The orbits of all Solar System planets and asteroids are ellipses.

Escape velocity: the minimum speed needed for an object to overcome the gravitational pull of the celestial body to which it is attracted.

Fall: a meteorite recovered after being observed in the process of falling to Earth.

Family: in astronomy, a group of asteroids that share similar orbits, eccentricities, and inclinations but do not necessarily travel in a cluster. A family may contain more than 200 members, believed to be fragments of a shattered parent body.

Find: a meteorite whose discovery did not stem from observations of its fall.

Fireball: an exceptionally bright meteor.

Fluorescence: a process in which the molecules of gas in a comet's coma absorb sunlight and radiate light themselves.

Fusion crust: the glassy exterior of a meteorite, formed by surface melting during the object's passage through Earth's atmosphere.

Goniometer: an optical instrument used to measure solid angles—for example, those between the faces of a crystal.

Greeks: a group of asteroids, named for Greek warriors, clustered at a point 60 degrees ahead of Jupiter on the planet's orbital path. *See* Trojans.

Half-life: the time required for half of the atoms in a sample of radioactive material to decay into a daughter element. Scientists use this value to calculate the ages of various meteorites.

Hydrogen cloud: a diffuse region of hydrogen atoms surrounding the nucleus of a comet and extending millions of miles into space. Unlike the comet's coma, or head, the hydrogen cloud is visible only in ultraviolet light.

Hydroxyl: a molecule, OH, consisting of one oxygen atom and one hydrogen atom, produced when a molecule of water (H_2O) is deprived of a hydrogen atom.

Impact crater: a large, bowl-shaped depression in Earth's surface, made when a large meteoroid or a comet strikes the planet.

Inclination: the angle between the orbit of an object such as an asteroid and the ecliptic plane.

Inclusion: a pocket of solid, liquid, or gaseous material whose chemical composition differs from that of the mineral or rock in which it is encased.

Inner cloud: one of three hypothesized cometary reservoirs, extending 30 to 10,000 astronomical units from the Sun. *See* Kuiper belt; Oort cloud.

Ion: an atom that has lost or gained one or more electrons, thus becoming electrically charged. By contrast, a neutral atom has an equal number of negatively charged electrons and positively charged protons, giving the atom a zero net electrical charge.

Ionopause: an area inside the contact surface of a comet where the solar wind cannot penetrate and only cometary matter exists.

Iridium: an exceptionally corrosion-resistant, hard, brittle, whitish yellow metallic element found in platinum ores. Though rare on Earth, it is abundant in meteorites.

Isotope: one of two or more atomic variants of a chemical element, having the same number of protons as the most common form of the element but a differing number of neutrons. Because they have unstable atomic nuclei, radioactive isotopes emit radiation spontaneously.

Jet: a plume of gas and dust that bursts from the nucleus of a comet as the comet approaches the Sun. Jets erupt from spots on the heated nucleus where the crust is relatively thin or the nuclear ice and dust are relatively loose. Jets can alter a comet's orbit; they also release the substances that form the comet's head and tails.

Kamacite: a nickel-iron alloy containing a relatively low proportion (up to 7.5 percent by weight) of nickel.

Kelvin: the name given to the temperature scale in which 0 denotes absolute zero (-273 degrees Celsius) and a unit of temperature, called a Kelvin, equals one Celsius degree.

Kirkwood gaps: orbital voids in the main asteroid belt that have been cleared out by the gravitational influence of Jupiter. They are named for their discoverer, American astronomer-mathematician Daniel Kirkwood.

Kuiper belt: the closest of three postulated cometary reservoirs, lying just beyond Neptune's orbit. It is named after the American astronomer Gerard Kuiper, who first suggested its existence. *See* Inner cloud; Oort cloud.

Lagrangian points: regions of gravitational stability in a planet's orbit where smaller bodies can travel in equilibrium with the planet. Jupiter's Lagrangian points, for example, harbor hundreds of asteroids, two-thirds of them at the point 60 degrees ahead of the planet (the Greeks) and the others at the point 60 degrees behind it (the Trojans). The points are named for Joseph Lagrange, the French mathematician who first posited their existence.

Long-period comet: a comet that requires 100,000 to 1,000,000 years to orbit the Sun, so that thousands of years may elapse between successive visits to the inner Solar System.

Magma: molten rocky matter, including dissolved gases and suspended crystals, that forms beneath the surface of a planet or an asteroid.

Magnetic field: a field of force around the Sun and each planet, generated by electrical currents, in which a magnetic influence is felt by other currents. The Sun's magnetic field, like that of the planets, exhibits a north and south pole linked by lines of varying magnetic strength and direction.

Magnetometer: a device for measuring the strength and direction of a magnetic field.

Magnetosheath: the turbulent region, between a comet's bow shock and its ionopause, where the solar wind is slowed and cooled by cometary ions as it approaches the comet's nucleus.

Mantle: the layer of a planet or asteroid that separates the core from the outermost layer, what is known as the crust. Among the asteroids, only those that are differentiated contain a mantle. *See* Differentiation.

Mass: a measure of the total amount of material in an object, determined by the object's gravity or by its tendency to resist acceleration.

Mass spectrometer: a device used to determine the chemical composition of a substance by measuring the varying masses of the substance's components.

Meteor: a streak of light in the sky created by the passage through Earth's atmosphere of a small rock or a piece of dust; also, the luminous object itself.

Meteorite: the recovered fragment of a meteoroid that has survived its transit through Earth's atmosphere. The weight of a meteorite may range from just a few ounces to nearly a hundred tons.

Meteoroid: a small metallic or rocky body orbiting the Sun in interplanetary space. A meteoroid that becomes visible as a streak of light upon entering Earth's atmosphere is called a meteor; meteoroids that survive their atmospheric passage and reach the planet's surface are known as meteorites.

Nebula: a cloud of interstellar dust or gas, in some cases a supernova remnant or a shell ejected by a star.

Neutron: an uncharged particle with a mass similar to that of a proton; normally found in an atom's nucleus.

Occultation: the partial blocking or eclipsing of one celestial body by another. By measuring the duration and extent of the eclipse, astronomers can calculate the diameter of the occulting object.

Oort cloud: in astronomical theory, the largest and most distant of three contiguous cometary reservoirs. Named for its proposer, Dutch astronomer Jan Oort, it is envisioned as a huge, spherical cloud that surrounds the Solar System and extends some ten trillion miles from the Sun. *See* Inner cloud; Kuiper belt.

Parabola: a line connecting the set of points in a plane that lie at an equal distance from both a fixed line and a fixed point in the same plane or a parallel plane.

Parallax: the apparent motion of a heavenly body on the celestial sphere. It results from the motion of a single observer or from the difference in location of two or more observers.

Particle detector: an instrument used to record the density, energy, or composition of such fundamental components of matter as molecules, atoms, protons, neutrons, and electrons.

Perihelion: the point in the orbit of a celestial object when it is nearest the Sun.

Photodissociation: the breaking down of molecules by solar photons. It is one of three processes that transform cometary gases as they leave the comet's nucleus. *See* Charge exchange; Photoionization.

Photoionization: the absorption of solar photons by neutral atoms, often yielding ions that form a comet's straight blue gas tail. *See* Charge exchange; Photodissociation.

Photometry: the measurement of an object's brightness, or apparent magnitude, by means of an instrument that counts the number of photons detected in a given time period.

Photon: a packet of electromagnetic energy that behaves like a chargeless particle and travels at the speed of light.

Planetesimal: in astronomical theory, a small, primitive body that orbited the Sun in the solar nebula, gaining mass through random collisions with other orbiting bodies until it eventually became a full-scale planet. The term derives from "infinitesimal planet."

Plasma: a gaslike conglomeration of charged particles that respond collectively to electrical currents and magnetic fields. Considered a fourth state of matter (along with solids, liquids, and gases), plasma constitutes the bulk of the Sun and the solar wind.

Polarity: the intrinsic polar orientation or alignment of a force such as the magnetic field of a star or planet.

Prograde motion: the orbital movement of a group of celestial bodies—the nine planets of the Solar System, for example—in one prevailing direction. Most comets show prograde motion; that is, they orbit the Sun in the same direction Earth does. *See* Retrograde motion.

Proton: a positively charged particle, normally found in an atom's nucleus, with 1,836 times the mass of an electron.

Regolith: a layer of fragmented rocky debris that forms the surface of many asteroids. It is produced by collisions among the asteroids.

Resonance: the enhanced gravitational effect on a small body when its orbital period is a simple fraction of a larger neighbor's. *See* Kirkwood gaps.

Retrograde motion: the real or apparent motion of a celestial body against the prevailing direction of movement of other bodies. The seemingly backward motion of the outer planets as they are overtaken by Earth is an example of apparent retrograde motion; the flight of Halley's comet, by contrast, is an example of real retrograde motion, because the comet orbits the Sun in a direction opposite Earth's.

Short-period comet: a comet whose solar orbit extends only as far as the outer planets and thus requires less than 200 years to complete.

Silica: a colorless or white chemical crystalline compound, also called silicon dioxide, that occurs in terrestrial sand and quartz, as well as in many stony meteorites.

Silicate: any of the largest and most common class of minerals, based on the nonmetallic element silicon. Nearly all rock-forming minerals are silicates, consisting of metal combined with silicon and oxygen.

Solar nebula: the diffuse primordial matter believed to have surrounded the young Sun and to have formed, by accretion, the nine planets as well as all comets, asteroids, and meteoroids.

Solar wind: a continuous current of charged particles that streams outward from the Sun through the Solar System.

Spectrograph: an instrument that splits light or other electromagnetic radiation into its individual wavelengths, collectively known as a spectrum, and records the result photographically or electronically. A spectrograph that lacks such a recording capability is called a spectroscope.

Spectrometer: a spectroscope that has been fitted with scales to measure the position of various spectral lines.

Spectrophotometry: the process of measuring the wavelengths of light reflected by an asteroid or other celestial body in certain regions of the spectrum.

Spectroscope: any of various instruments used for the direct observation of a spectrum.

Spectroscopy: the study of spectra, including the position and intensity of emission and absorption lines, to determine the chemical elements or physical processes that created them.

Spectrum: the array of electromagnetic radiation, arranged in order of wavelength from long-wave radio emissions to short-wave gamma rays; also, a narrower band of wavelengths, called the visible spectrum, as when light dispersed by a prism shows its component colors. Spectra are often striped with emission or absorption lines, which can be examined to reveal the composition and motion of the light source.

Spherules: tiny, metal-rich, glassy spheres whose high nickel content indicates that they are of extraterrestrial origin.

Stishovite: a mineral that is produced when the shock wave from a particularly high-pressure meteorite impact passes through rock containing quartz. *See* Coesite.

Strewn field: the natural distribution on the ground of meteoritic debris, generally in the form of an ellipse, that results from the fragmentation of a meteoroid during its atmospheric transit.

Terminator: the line dividing day and night on the surface of a planet, moon, comet, or asteroid.

Tidal force: a gravitational pull exerted by a star, planetary system, or galaxy, whose strength or direction tends to alter the orbit or shape of another celestial object.

Torus: a doughnut-shaped volume of space, such as that circumscribed by the orbits of the asteroids in the main belt between Mars and Jupiter.

Trojans: a group of asteroids, named for Trojan warriors, clustered at a point 60 degrees behind Jupiter on the planet's orbital path. Also, the collective term for this group and a second group, clustered 60 degrees ahead of Jupiter. *See* Greeks.

Troposphere: the lowest level of Earth's atmosphere, extending about ten miles above the surface of the planet.

Vaporization: the sudden conversion of a solid or liquid into a gaseous state.

Widmannstätten pattern: the regular, geometric array of the crystals in certain types of iron meteorites that is revealed by etching a sample of the meteorite with a dilute solution of acid.

BIBLIOGRAPHY

Books

Abell, George Ogden. *Exploration of the Universe* (5th ed.). New York: CBS College Publishing, 1987.

Asimov, Isaac. *Asimov on Astronomy.* Garden City, N.Y.: Anchor Press/Doubleday, 1975.

Beatty, J. Kelly, Brian O'Leary, and Andrew Chaikin (eds.). *The New Solar System* (2d ed.). Cambridge, Mass.: Sky, 1981.

Binzel, Richard P., Tom Gehrels, and Mildred Shapley Matthews (eds.). *Asteroids II.* Tucson: University of Arizona Press, 1989.

Brandt, John C. *Comets: Readings from Scientific American.* San Francisco: W. H. Freeman, 1981.

Briggs, John, and F. David Peat. *Turbulent Mirror: An Illustrated Guide to Chaos Theory and the Science of Wholeness.* New York: Harper & Row, 1989.

Brown, Peter Lancaster. *Comets, Meteorites and Men.* New York: Taplinger, 1973.

Burke, John G. *Cosmic Debris: Meteorites in History.* Berkeley: University of California Press, 1986.

Calder, Nigel. *The Comet Is Coming!* New York: Viking Press, 1980.

Carr, Michael H. *The Surface of Mars.* New Haven, Conn.: Yale University Press, 1981.

Chapman, Clark R., and David Morrison. *Cosmic Catastrophes.* New York: Plenum Press, 1989.

Chapman, Robert D., and John C. Brandt. *The Comet Book: A Guide for the Return of Halley's Comet.* Boston: Jones and Bartlett, 1984.

Cunningham, Clifford J. *Introduction to Asteroids.* Richmond, Va.: Willmann-Bell, 1988.

Darling, David J. *Comets, Meteors, and Asteroids: Rocks in Space.* Minneapolis, Minn.: Dillon Press, 1984.

Davies, John Keith. *Cosmic Impact.* New York: St. Martin's Press, 1986.

Debus, Allen G. (ed.). *World Who's Who in Science.* Chicago: A. N. Marquis, 1968.

Dodd, Robert T. *Thunderstones and Shooting Stars: The Meaning of Meteorites.* Cambridge, Mass.: Harvard University Press, 1986.

Freitag, Ruth S. *Halley's Comet: A Bibliography.* New York: Taplinger, 1973.

Gehrels, Tom (ed.). *Asteroids.* Tucson: University of Arizona Press, 1979.

Glass, Billy P. *Introduction to Planetary Geology.* Cambridge, England: Cambridge University Press, 1982.

Graham, A. L., A. W. R. Bevan, and R. Hutchison. *Catalogue of Meteorites* (4th ed.). Tucson: University of Arizona Press, 1985.

Green, Jay. *McGraw-Hill Modern Men of Science.* New York: McGraw-Hill, 1966.

Hutchison, Robert. *The Search for Our Beginning.* New York: Oxford University Press, 1983.

Jackson, Joseph H. *Pictorial Guide to the Planets* (3d ed.). New York: Harper & Row, 1981.

Kerridge, J. F., and M. S. Matthews (eds.). *Meteorites and the Early Solar System.* Tucson: University of Arizona Press, 1988.

Knight, David C. *The Tiny Planets: Asteroids of Our Solar System.* New York: William Morrow, 1973.

Kowal, Charles T. *Asteroids: Their Nature and Utilization.* New York: Halsted Press, 1988.

LeMaire, T. R. *Stones from the Stars: The Unsolved Mysteries of Meteorites.* Englewood Cliffs, N.J.: Prentice-Hall, 1980.

Littmann, Mark, and Donald K. Yeomans. *Comet Halley: Once in a Lifetime.* Washington, D.C.: American Chemical Society, 1985.

McSween, Harry Y., Jr. *Meteorites and Their Parent Planets.* Cambridge, England: Cambridge University Press, 1987.

Maran, Stephen. "Secrets from the Fallen Stars." In *1985 Yearbook of Science and the Future.* Chicago: Encyclopaedia Britannica, 1984.

Middlehurst, Barbara M., and Gerard P. Kuiper (eds.). *The Moon, Meteorites, and Comets.* Chicago: University of Chicago Press, 1963.

Moore, Carleton B. *Cosmic Debris.* Boston: Little, Brown, 1969.

Moore, Patrick (ed.). *The International Encyclopedia of Astronomy.* New York: Orion Books, 1987.

Nininger, H. H.:
  Arizona's Meteorite Crater. Denver, Colo.: American Meteorite Laboratory, 1956.
  Find a Falling Star. New York: Paul S. Eriksson, 1972.

The Rand McNally Atlas of the Oceans. New York: Rand McNally, 1977.

Raup, David M. *The Nemesis Affair: A Story of the Death of Dinosaurs and the Ways of Science.* Ontario: Penguin Books Canada, 1986.

Sagan, Carl, and Ann Druyan. *Comet.* New York: Random House, 1985.

Sears, D. W. *The Nature and Origin of Meteorites.* New York: Oxford University Press, 1978.

Space Travel and Astronomy. New York: Crescent Books, 1979.

Stewart, Ian. *Does God Play Dice? The Mathematics of Chaos.* New York: Basil Blackwell, 1989.

Watson, Fletcher G. *Between the Planets.* Cambridge, Mass.: Harvard University Press, 1956.

Whipple, Fred L. *The Mystery of Comets.* Washington, D.C.: Smithsonian Institution Press, 1985.

Periodicals

Ahrens, Thomas J., and John D. O'Keefe. "Impact on the Earth, Ocean and Atmosphere." *International Journal of Impact Engineering*, 1987, Vol. 5, pp. 13–32.

Alvarez, Luis W. "Mass Extinctions Caused by Large Bolide Impacts." *Physics Today*, July 1987.

Alvarez, Walter, et al. "Impact Theory of Mass Extinctions and the Invertebrate Fossil Record." *Science*, March 16, 1984.

"AstroNews: Earth Gains a Neighbor." *Astronomy*, August 1989.

Bagnall, Philip M. "The Quadrantids: Bright Prospects for 1987." *Astronomy*, January 1987.

Balsiger, Hans, Hugo Fechtig, and Johannes Geiss. "A Close Look at Halley's Comet." *Scientific American*, September 1988.

Bassett, Carol Ann. "If an Asteroid Threatens Earth, Tom Gehrels Will Sound the Alarm." *People*, September 19, 1984.

Beatty, J. Kelly:
  "The High Road to Halley." *Sky & Telescope*, March 1986.
  "An Inside Look at Halley's Comet." *Sky & Telescope*, May 1986.

Berry, Richard. "Search for the Primitive." *Astronomy*, June 1987.

Bogard, Donald. "A Meteorite from the Moon." *Geophysical Research Letters*, September 1983.

Bohor, Bruce F., Peter J. Modreski, and Eugene E. Foord. "Shocked Quartz in the Cretaceous-Tertiary Boundary Clays: Evidence for a Global Distribution." *Science*, May 8, 1987.

Bosveld, Jane. "Apocalypse, How? Stars." *Omni*, December 1989.

Brown, John F., Jr., et al. "Polychlorinated Biphenyl Dechlorination in Aquatic Sediments." *Science*, May 8, 1987.

Cassidy, William A., and Louis A. Rancitelli. "Antarctic Meteorites." *American Scientist*, March-April 1982.

Chapman, Clark R. "The Nature of Asteroids." *Scientific American*, January 1975.

Chapman, Clark R. and David Morrison:
"Cosmic Impacts, Cosmic Catastrophes." *Mercury*, November-December 1989.
"The Next Doomsday Impact." *Astronomy*, November 1989.

Crutchfield, James P., et al. "Chaos." *Scientific American*, December 1986.

Davis, D. R., et al. (eds.). *Journal of the Italian Astronomical Society: Catastrophic Disruption of Asteroids and Satellites*, 1986, Vol. 57, no. 1.

Delsemme, Armand H. "Whence Come Comets?" *Sky & Telescope*, March 1989.

Dodd, Robert T. "Unique Find from Antarctica." *Nature*, March 23, 1989.

Eberhart, J.:
"Comet Halley Encounters Earth's Space Age." *Science News*, March 22, 1986.
"Radar Reveals an Asteroid's Strange Shape." *Science News*, November 25, 1989.

Eicher, David J., "Last Look at Halley." *Astronomy*, September 1986.

Eugster, O. "History of Meteorites from the Moon Collected in Antarctica." *Science*, September 15, 1989.

"Getting Mars' Rocks Off." *Discovery*, December 1987.

Gleick, James. "A Handful of Mysterious Meteorites Bring the Study of Mars to Earth." *New York Times*, September 1, 1987.

Gore, Rick. "What Caused Earth's Great Dyings?" *National Geographic*, June 1989.

"IRAS Asteroid Catalogue." *Sky & Telescope*, June 1987.

Kaneda, E., et al. "Observation of Comet Halley by the Ultraviolet Imager of Suisei." *Nature*, May 15, 1986.

Kerr, Richard A.:
"The Great Asteroid Roast: Was It Rare or Well-Done?" *Science*, February 2, 1990.
"An Impact but No Volcano." *Science*, May 25, 1984.
"Martian Meteorites Are Arriving." *Science*, August 14, 1987.
"Periodic Extinctions and Impacts Challenged." *Science*, March 22, 1985.
"S Asteroids at Controversy's Core." *Science*, February 2, 1990.

Knacke, Roger. "Sampling the Stuff of a Comet." *Sky & Telescope*, March 1987.

"Largest Radar Detects Dumbbell in Space." *Science*, November 24, 1989.

Lewis, Roy S., and Edward Anders. "Interstellar Matter in Meteorites." *Scientific American*, August 1983.

MacDougall, J. D. "Seawater Strontium Isotopes, Acid Rain, and the Cretaceous-Tertiary Boundary." *Science*, January 29, 1988.

MacRobert, Alan M. "Backyard Astronomy." *Sky & Telescope*, August 1988.

Maran, Stephen P.:
"Gaps in the Asteroid Belt." *Natural History*, August 1986.
"Meteorites—How Science Reads Them." *Popular Science*, April 1980.
"A Near Miss." *Natural History*, March 1981.
"Rocks from Mars." *Natural History*, November 1983.
"What Struck Tunguska?" *Natural History*, February 1984.

Marvin, Ursula G. "The Discovery and Initial Characterization of Allan Hills 81005: The First Lunar Meteorite." *Geophysical Research Letters*, September 1983.

"Massive Antarctic Meteorite Found." *Sky & Telescope*, June 1989.

Melosh, H. J. "Ejection of Rock Fragments from Planetary Bodies." *Geology*, February 1985.

Melosh, H. J., and A. M. Vickery. "The Large Crater Origin of SNC Meteorites." *Science*, August 14, 1987.

Monastersky, R. "Heat Wave at the K-T Boundary?" *Science News*, March 12, 1988.

Moorbath, Stephen. "The Most Ancient Rocks Revisited." *Nature*, June 19, 1986.

Morrison, David, and Clark R. Chapman. "Target Earth: It Will Happen." *Sky & Telescope*, March 1990.

"Need One More Thing to Worry About?" *Newsweek*, May 1, 1989.

"News Notes: Sing the Asteroid Electric." *Sky & Telescope*, April 1989.

Nishiizumi, K., et al. "Age of Allan Hills 82102, a Meteorite Found Inside the Ice." *Nature*, August 17, 1989.

O'Keefe, John D., and Thomas J. Ahrens:
"Cometary and Meteorite Swarm Impact on Planetary Surfaces." *Journal of Geophysical Research*, August 10, 1982.
"Impact Mechanics of the Cretaceous-Tertiary Extinction Bolide." *Nature*, July 8, 1982.

Peterson, I.:
"Earthward on a Rocky, Chaotic Course." *Science News*, July 13, 1985.
"Star Dust in the Sky with Diamonds." *Science News*, March 14, 1987.
"Unexpected Asteroid: A Close Call from Space." *Science News*, May 6, 1989.

Pool, Robert. "Chaos Theory: How Big an Advance?" *Science*, July 7, 1989.

Prinn, Ronald G., and Bruce Fegley, Jr. "Bolide Impacts, Acid Rain, and Biospheric Traumas at the Cretaceous-Tertiary Boundary." *Earth and Planetary Science Letters*, 1987, Vol. 83, pp. 1-15.

"Radar Imaging Captures Exotic Asteroid." *Sky & Telescope*, January 1990.

Rampino, Michael. "Dinosaurs, Comets and Volcanoes." *New Scientist*, February 18, 1989.

Reitsema, H. J., W. A. Delamere, and F. L. Whipple. "Active Polar Region on the Nucleus of Comet Halley." *Science*, January 13, 1989.

Robertson, Donald Frederick. "Return to Mars: Soviet Phobos Mission to Probe Moons of Mars." *Astronomy*, November 1987.

Rogers, Michael. "The Death of the Dinosaurs." *Newsweek*, December 19, 1988.

Sagdeev, Roald Z., and A. Galeev. "Comet Halley and the Solar Wind." *Sky & Telescope*, March 1987.

Sagdeev, Roald Z., and Leonid V. Ksanfomality. "Half an Hour in the Comet's Coma." *Planetary Report*, September-October 1987.

Schramm, David N., and Robert N. Clayton. "Did a Supernova Trigger the Formation of the Solar System?" *Scientific American*, October 1978.

Schwarzschild, Bertram M. "Chaotic Orbits and Spins in the Solar System." *Physics Today*, September 1985.

Sinnott, Roger W. "An Asteroid Whizzes Past Earth." *Sky & Telescope*, July 1989.

"Soviet Space Odyssey." *Sky & Telescope,* October 1985.

Spratt, Christopher E. "On the Trail of a Meteorite." *Astronomy,* August 1989.

"Star-Struck?" *Scientific American,* April 1988.

Thomsen, D. E. "Signs of Nemesis: Meteors, Magnetism." *Science News,* February 14, 1987.

Trefil, James. "Stop to Consider the Stones that Fall from the Sky." *Smithsonian,* September 1989.

Tsung, Thomas. "The Jilin Meteorite." *Sky & Telescope,* June 1978.

Van Den Bergh, Sidney. "Life and Death in the Inner Solar System." *Publications of the Astronomical Society of the Pacific,* May 1989.

Waldrop, M. Mitchell. "After the Fall." *Science,* February 26, 1988.

Weaver, Kenneth F. "Meteorites: Invaders from Space." *National Geographic,* September 1986.

Weissman, Paul R. "Realm of the Comets." *Sky & Telescope,* March 1987.

Wetherill, George W. "The Allende Meteorite." *Natural History,* November 1978.

"Where Have All the Dinos Gone?" *U.S. News & World Report,* July 6, 1987.

Whipple, Fred L. "The Black Heart of Comet Halley." *Sky & Telescope,* March 1987.

Wright, I. P., M. M. Grady, and C. T. Pillinger. "Organic Materials in a Martian Meteorite." *Nature,* July 20, 1989.

Other Sources

Cleggett-Haleim, Paula. "NASA Astronomer Discovers 'Near Miss' Asteroid That Passed Earth." News release. Washington, D.C.: NASA, April 19, 1989.

Jones, Eric M., and John W. Kodis. "Atmospheric Effects of Large Body Impacts: The First Few Minutes." Conference paper in Special Paper 190. Boulder, Colo.: Geological Society of America, 1982.

O'Keefe, John D. "The Interaction of the Cretaceous/Tertiary Extinction Bolide with the Atmosphere, Ocean, and Solid Earth." Conference paper in Special Paper 190. Boulder, Colo.: Geological Society of America, 1982.

Reitsema, H. J., et al. "Nucleus Morphology of Comet Halley." Paper presented at the 20th ESLAB Symposium on the Exploration of Halley's Comet, Heidelberg, West Germany, October 27-31, 1986.

Shoemaker, Eugene M., Carolyn Shoemaker, and Ruth F. Wolfe. "Asteroid and Comet Flux in the Neighborhood of the Earth." Paper presented at "Global Catastrophes in Earth History: An Interdisciplinary Conference on Impacts, Volcanism, and Mass Mortality," Snowbird, Utah, October 20-23, 1988.

Silver, Leon T., and Peter H. Schultz (eds.). "Geological Implications of Impacts of Large Asteroids and Comets on the Earth." Complete conference proceedings, Special Paper 190. Boulder, Colo.: Geological Society of America, 1982.

Stiles, Lore. "Asteroid Impact Would Have Lit Global Wildfires, Broiled Dinosaurs Alive, Scientists Report in *Nature.*" News release. Tucson: University of Arizona, January 18, 1990.

Toon, O. B., et al. "Evolution of an Impact-Generated Dust Cloud and Its Effects on the Atmosphere." Conference paper in Special Paper 190. Boulder, Colo.: Geological Society of America, 1982.

INDEX

ACKNOWLEDGMENTS

The editors wish to thank Roy S. Clarke, Smithsonian Institution, Washington, D.C.; Armand Delsemme, Toledo, Ohio; Billy P. Glass, University of Delaware, Newark; James W. Head, Brown University, Providence, R.I.; Eleanor F. Helin, California Institute of Technology, Pasadena; Jack Hills, Los Alamos National Laboratory, Los Alamos, N.Mex.; Glenn I. Huss and Margaret Huss, American Meteorite Laboratory, Denver, Colo.; Tom Ligon, "Dance of the Planets"/ARC Software, Inc., Loveland, Colo.; Derek Sears, University of Arkansas, Fayetteville; Eugene M. Shoemaker, U.S. Geological Survey, Flagstaff, Ariz.; Gerald J. Wasserburg, California Institute of Technology, Lunar Research Laboratory, Pasadena; Paul Weissman, Jet Propulsion Laboratory, Pasadena, Calif.; Richard West, European Southern Observatory, Garching, West Germany; David Winningham, Southwest Research Institute, San Antonio, Tex.; and John Wood, Harvard-Smithsonian Center for Astrophysics, Cambridge, Mass.

PICTURE CREDITS

The sources for the illustrations in this book are listed below. Credits from left to right are separated by semicolons; credits from top to bottom are separated by dashes.

Cover: Art by Stephen R. Wagner. 6, 7: Royal Observatory, Edinburgh. 8: Initial cap, detail from pages 6, 7. 10: Beijing Cultural Relics Publishing House. 11: Giraudon, Paris. 12, 13: U.S. Naval Observatory. 15-19: Background art by Fred Holz and Stephen Wagner. 15: Royal Society, London; Yerkes Observatory, Williams Bay, Wis. 16, 17: Musée de la Poste, Paris; courtesy Donald K. Yeomans; drawings by George Bond, from "An Account of Donati's Comet," Harvard Observatory, 1858, courtesy Donald K. Yeomans; Larry Sherer, courtesy U.S. Naval Observatory; Institute of Astronomy, University of Cambridge; courtesy Donald K. Yeomans. 18, 19: Yerkes Observatory; JPL/Donald K. Yeomans; Smithsonian Astrophysical Observatory; drawings by George Bond, from "An Account of Donati's Comet," Harvard Observatory, 1858, courtesy Donald K. Yeomans; Glasheen Graphics, La Jolla, Calif. 21: S. G. Djorgovski, taken at Kitt Peak National Observatory—Larry Sherer, from *The Comets and Their Origin*, by R. A. Lyttleton, Cambridge University Press, London, 1953. 22-27: Art by Alfred T. Kamajian. 29: Copied by Larry Sherer, from *Mystery of Comets*, by Fred L. Whipple, Smithsonian Institution Press, Washington, D.C., 1985—drawings by George Bond, from "An Account of Donati's Comet," Harvard Observatory, 1858 (2). 30: Larry Sherer, courtesy U.S. Naval Observatory. 32-35: Art by Stephen R. Wagner. 36: Naval Research Laboratory. 38, 39: Peter Stättmayer, Treugesell-Verlag, Düsseldorf; New Mexico State University Observatory, Las Cruces (4). 40, 41: Art by Matt McMullen. Background art by Time-Life Books. 42, 43: Art by Rob Wood of Stansbury, Ronsaville, Wood, Inc. 44, 45: Art by Yvonne Gensurowsky of Stansbury, Ronsaville, Wood, Inc. 46, 47: Institute of Space and Astronautical Science, Ministry of Education, Science and Culture, Tokyo; J. D. Winningham, Southwest Research Institute, San Antonio, Tex.; Institute of Space and Astronautical Science, Ministry of Education, Science and Culture, Tokyo. All art by Yvonne Gensurowsky of Stansbury, Ronsaville, Wood, Inc. 48: Courtesy Harold Reitsema, Ball Aerospace Systems Division, © Max Planck Institut für Radioastronomie, Bonn; courtesy J. A. Simpson, "The U.S.S.R. Spacecraft Visits to Comet Halley: The Missions and the Scientific Results," *Proceedings of the American Philosophical Society*, Special Publication APS no. 41, 1987; © 1986 Space Research Institute—courtesy Alan Delamere and Harold Reitsema, Ball Aerospace Systems Division, © Max Planck Institut für Radioastronomie, Bonn. 49: Art by Rob Wood of Stansbury, Ronsaville, Wood, Inc. 50, 51: From "Dance of the Planets," A.R.C. Inc. Software, Loveland, Colo. 52, 53: Initial cap, detail from pages 50, 51. 54, 55: Art by Time-Life Books, based on information from A.R.C. Inc. Software, Loveland, Colo. 56, 57: Mary Lea Shane Archives of the Lick Observatory, Santa Cruz, Calif. (2); Ernst Zinner Astronomy Collection of San Diego State University Library/Special Collections (2); Mary Lea Shane Archives of the Lick Observatory, Santa Cruz, Calif. (2); Ernst Zinner Astronomy Collection of San Diego State University Library/Special Collections. Rock art by Fred Holz. 58, 59: Mary Lea Shane Archives of the Lick Observatory, Santa Cruz, Calif. (2); Koseisha Koseikaku Company, Ltd., Tokyo; Ernst Zinner Astronomy Collection of San Diego State University Library/Special Collections (2); Roger Ressmeyer-Starlight, South Hampton, N.Y.; courtesy Eleanor F. Helin and Palomar Observatory; Mark Sennet/Onyx, Los Angeles. Rock art by Fred Holz. 61: California Institute of Technology, courtesy Eleanor F. Helin. 62, 63: Art by Stephen R. Wagner. 64, 65: Painting by Andrew Chaiken; inset art by Time-Life Books. 66: From "Dance of the Planets," A.R.C. Inc. Software, Loveland, Colo. 68: Courtesy Glenn J. Veeder, JPL. 69: Art by Fred Holz. 70: Space Research Institute, USSR Academy of Sciences. 73-76: JPL/NASA photographs courtesy Steven Ostro, John Chandler, Alice Hine, Keith Rosema, Irwin Shapiro, and Donald K. Yeomans. 79-87: Art by Stephen R. Wagner. 88, 89: Photo Researchers, Inc., New York. 90: Initial cap, detail from pages 88, 89. 93: Jerry Wasserburg, Caltech Lunar Research Labs, Pasadena, Calif. 94, 95: Courtesy Department of Library Services, American Museum of Natural History, New York; Smithsonian Institution (2). 96: Smithsonian Institution; from *Stones from the Stars: The Unsolved Mysteries of Meteorites*, by T. R. LeMaire, Prentice-Hall, Englewood Cliffs, N.J., 1980—American Meteorite Laboratory, Denver. 98, 99: Smithsonian Institution; except bottom right, Photo Researchers, Inc., New York. 100: Smithsonian Astrophysical Observatory. 101: Courtesy Billy P. Glass. 102, 103: Photo Researchers, Inc., New York. 104, 105: University of Arkansas, Cosmochemistry Group, Fayetteville (2); Bettmann Archive, New York; Smithsonian Institution; American Meteorite Laboratory, Denver. Background photo courtesy Comstock, New York. 106: Smithsonian Institution. 111: Art by Matt McMullen—NASA, Washington, D.C. 112-115: Art by Matt McMullen, except page 115, upper right, by Time-Life Books. 117, 118: Art by Fred Holz. 119: Art by Fred Holz; Glen Izett, U.S. Geological Survey (2)—courtesy Walter Alvarez. 121-123: Larry Sherer, art by Bryn Barnard. 124, 125: Larry Sherer, art by Bryn Barnard, inset art by Lili Robins. 126, 127: Larry Sherer, art by Bryn Barnard, inset art by Lili Robins. 128-133: Larry Sherer, art by Bryn Barnard.

Time-Life Books
is a wholly owned subsidiary of
THE TIME INC. BOOK COMPANY

President and Chief Executive Officer:
Kelso F. Sutton
President, Time Inc. Books Direct:
Christopher T. Linen

TIME-LIFE BOOKS INC.
EDITOR: George Constable
Director of Design: Louis Klein
Director of Editorial Resources: Phyllis K. Wise
Director of Photography and Research:
John Conrad Weiser

PRESIDENT: John M. Fahey, Jr.
Senior Vice Presidents: Robert M. DeSena, Paul R.
Stewart, Curtis G. Viebranz, Joseph J. Ward
Vice Presidents: Stephen L. Bair, Bonita L.
Boezeman, Mary P. Donohoe, Stephen L.
Goldstein, Juanita T. James, Andrew P. Kaplan,
Trevor Lunn, Susan J. Maruyama,
Robert H. Smith
New Product Development: Trevor Lunn,
Donia Ann Steele
Supervisor of Quality Control: James King

PUBLISHER: Joseph J. Ward

Editorial Operations
Production: Celia Beattie
Library: Louise D. Forstall

Computer Composition: Gordon E. Buck
(Manager), Deborah G. Tait, Monika D. Thayer,
Janet Barnes Syring, Lillian Daniels

Correspondents: Elisabeth Kraemer-Singh (Bonn),
Christine Hinze (London), Christina Lieberman
(New York), Maria Vincenza Aloisi (Paris), Ann
Natanson (Rome). Valuable assistance was also
provided by Dick Berry (Tokyo), Elizabeth Brown
(New York), John Dunn (Melbourne), Barbara
Hicks (London), Kalypso Papdopoulou (London),
Bing Wong (Hong Kong).

VOYAGE THROUGH THE UNIVERSE

SERIES DIRECTOR: Roberta Conlan
Series Administrator: Susan Stuck

Editorial Staff for *Comets, Asteroids,
& Meteorites*
Designers: Cynthia T. Richardson (principal),
Ellen Robling
Associate Editor: Tina McDowell (pictures)
Text Editors: Allan Fallow (principal),
Stephen Hyslop, Robert M. S. Somerville
Researchers: Katya Sharpe Cooke, Dan Kulpinski,
Edward O. Marshall, Patricia Mitchell,
Barbara Sause
Writers: Stephanie Lewis, Barbara Mallen
Assistant Designer: Brook Mowrey
Copy Coordinator: Darcie Conner Johnston
Picture Coordinator: Jennifer Iker
Editorial Assistant: Katie Mahaffey

Special Contributors: Andrew Chaikin,
Stephen Maran, Gina Maranto, Eliot Marshall,
Sally Stephens, Larry Thompson, Elizabeth Ward
(text); Adam Dennis, Edward Dixon, Jocelyn Lind-
say, Cheryl Pellerin, Eugenia Scharf (research);
Barbara L. Klein (index).

CONSULTANTS

THOMAS J. AHRENS is a professor of geophysics at
the California Institute of Technology and director
of the Lindhurst Laboratory of Experimental Geophysics. He specializes in shock compression and
impact cratering, including the effects of meteorite
impact on Earth's atmosphere and ocean.

JOHN C. BRANDT, an astronomer at the University
of Colorado, Boulder, studies cometary plasma tails
and their interaction with the solar wind.

CLARK R. CHAPMAN of the Planetary Science Institute in Tucson, Arizona, analyzes the potential of
small asteroids hitting the Earth. He is also part of
the imaging team of the NASA Galileo mission to
Jupiter, a flight that includes a flyby of two asteroids.

ROBERT T. DODD is professor of geology and planetary science at the State University of New York at
Stony Brook. He has studied and written about meteorites for almost thirty years.

CHARLES KOWAL is an operations astronomer at
the Space Telescope Science Institute in Baltimore.
He has discovered numereous comets and asteroids
as well as eighty-one supernovae and Chiron, an
object that is found between the orbits of Saturn
and Uranus.

H. JAY MELOSH is a professor of planetary science
at the Lunar and Planetary Laboratory at the University of Arizona in Tucson. His research is largely
devoted to the study of impact cratering and its
effects on Earth and other planets in the Solar System.

MALCOLM BOWEN NIEDNER is an astrophysicist
at NASA Goddard Space Flight Center in Greenbelt,
Maryland. An expert on the International Halley
Watch team, he specializes in the study of comet
tails.

DAVID M. RAUP, paleontologist and author, is a
professor in the Department of Geophysical Sciences at the University of Chicago; he examines the role
of mass extinction in the geologic past.

FRED LAWRENCE WHIPPLE, a retired professor of
astronomy at Harvard University and the director
emeritus of the Smithsonian Institution Astrophysical Observatory in Cambridge, Massachusetts, has
done pioneering work on comets. The observatory
at Mount Hopkins, Arizona, has been renamed in his
honor.

DONALD K. YEOMANS, a senior research astronomer at the Jet Propulsion Laboratory in Pasadena,
California, provided the position predictions for the
encounter of five international spacecraft with
comet Halley in 1986. He is also a radio science team
leader for CRAF (Comet Rendezvous Asteroid
Flyby), an international cooperative space mission
scheduled for launch in 1995.

**Library of Congress Cataloging in
Publication Data**
Comets, asteroids, and meteorites / by the editors
of Time-Life Books.
   p.   cm.   (Voyage through the universe).
Bibliography: p.
Includes index.
ISBN 0-8094-6904-9
ISBN 0-8094-6905-7 (lib. bdg.).
1. Comets. 2. Asteroids. 3. Meteorites.
I. Time-Life Books.
II. Title: Comets, asteroids, and meteorites.
III. Series.
QB721.C649   1990
523.6—dc20   90-10858 CIP

For information on and a full description of
any of the Time-Life Books series, please call
1-800-621-7026 or write:
Reader Information
Time-Life Customer Service
P.O. Box C-32068
Richmond, Virginia 23261-2068.

Time-Life Books Inc. offers a wide range of fine
recordings, including a *Rock 'n' Roll Era* series.
For subscription information, call 1-800-621-7026
or write Time-Life Music, P.O. Box C-32068, Richmond, Virginia 23261-2068.

Earth: diameter 7,926 miles

Neptune: diameter 30,700 miles

Uranus: diameter 31,600 miles

Red supergiant: diameter 400 million miles

Solar System: diameter 7.5 billion miles

Globular cluster: diameter 2×10^{14} miles

Milky Way: diameter 100,000 light-years

Local Group of galaxies:
6 million light-years across

Largest double radio source:
length 17 million light-years